T0192364

Handbook of Educational Measurement and Psychometrics Using R

Chapman & Hall/CRC
The R Series

Series Editors

Basics of Matrix Algebra for Statistics with R
Nick Fieller

Introductory Fisheries Analyses with R
Derek H. Ogle

Statistics in Toxicology Using R
Ludwig A. Hothorn

Spatial Microsimulation with R
Robin Lovelace, Morgane Dumont

Extending R
John M. Chambers

Using the R Commander: A Point-and-Click Interface for R
John Fox

Computational Actuarial Science with R
Arthur Charpentier

bookdown: Authoring Books and Technical Documents with R Markdown
Yihui Xie

Testing R Code
Richard Cotton

R Primer, Second Edition
Claus Thorn Ekstrøm

Flexible Regression and Smoothing: Using GAMLSS in R
Mikis D. Stasinopoulos, Robert A. Rigby, Gillian Z. Heller, Vlasios Voudouris, and Fernanda De Bastiani

The Essentials of Data Science: Knowledge Discovery Using R
Graham J. Williams

blogdown: Creating Websites with R Markdown
Yihui Xie, Alison Presmanes Hill, and Amber Thomas

For more information about this series, please visit: https://www.crcpress.com/go/the-r-series

Handbook of
Educational
Measurement
and Psychometrics
Using R

Christopher D. Desjardins
Okan Bulut

CRC Press
Taylor & Francis Group
Boca Raton London New York

CRC Press is an imprint of the
Taylor & Francis Group, an **informa** business
A CHAPMAN & HALL BOOK

CRC Press
Taylor & Francis Group
6000 Broken Sound Parkway NW, Suite 300
Boca Raton, FL 33487-2742

First issued in paperback 2020

ISBN-13: 978-1-4987-7013-2 (hbk)
ISBN-13: 978-0-367-73467-1 (pbk)

Library of Congress Cataloging-in-Publication Data

Names: Desjardins, Christopher David, author. | Bulut, Okan, author.
Title: Handbook of educational measurement and psychometrics using R : By Christopher David Desjardins, Okan Bulut.
Description: Boca Raton, Florida : CRC Press, [2018] | Includes bibliographical references and index.
Identifiers: LCCN 2017057935| ISBN 9781498770132 (hardback : alk. paper) | ISBN 9781315154268 (e-book : alk. paper) | ISBN 9781498770149 (e-book (pdf) : alk. paper) | ISBN 9781351650304 (e-book (epub) : alk. paper) | ISBN 9781351640770 (e-book (mobi/kindle) : alk. paper)
Subjects: LCSH: Educational tests and measurements--Handbooks, manuals, etc. | Psychometrics--Methodology--Handbooks, manuals, etc. | R (Computer program language)--Handbooks, manuals, etc.
Classification: LCC LB3051 .D4465 2018 | DDC 371.26--dc23
LC record available at https://lccn.loc.gov/2017057935

Visit the Taylor & Francis Web site at
http://www.taylorandfrancis.com

and the CRC Press Web site at
http://www.crcpress.com

CDD: To Kirsten

OB: To my parents and my brother

Contents

Preface   xiii

List of Figures   xix

List of Tables   xxiii

1 **Introduction to the R Programming Language**   1
   1.1 Chapter Overview   1
   1.2 What Is R?   1
        1.2.1 Our Approach to R   2
   1.3 Obtaining and Installing R   3
        1.3.1 Windows   4
        1.3.2 Mac   4
        1.3.3 Linux   4
   1.4 Obtaining and Installing RStudio   4
   1.5 Using R   6
        1.5.1 Basic R Usage   7
        1.5.2 R Packages   11
               1.5.2.1 Masked Functions   13
        1.5.3 Assessing and Reading in Data   14
        1.5.4 Data Manipulation   16
        1.5.5 Descriptive and Inferential Statistics   21
        1.5.6 Plotting in R   25
               1.5.6.1 Base R Graphics   25
               1.5.6.2 Lattice Graphics   27
   1.6 Installing Packages Used in This Handbook   28
   1.7 Chapter Summary   29

2 **Classical Test Theory**   31
   2.1 Chapter Overview   31
   2.2 What Is Measurement?   31
   2.3 Issues in Measurement   32
        2.3.1 Type of Scales   33
   2.4 The Classical Test Theory Framework   40
        2.4.1 Reliability   41
        2.4.2 Validity   47
        2.4.3 Item Analysis   49

2.5   Summary   53

3   Generalizability Theory   **55**
3.1   Chapter Overview   55
3.2   Introduction   55
3.3   Examples   60
      3.3.1   One-Facet Design   60
              3.3.1.1   G Study   60
              3.3.1.2   D Study   64
      3.3.2   Two-Facet Crossed Design   66
              3.3.2.1   G Study   66
              3.3.2.2   D Study   68
      3.3.3   Two-Facet Partially Nested Design   69
              3.3.3.1   G Study   70
              3.3.3.2   D Study   71
      3.3.4   Two-Facet Crossed Design with a Fixed Facet   72
              3.3.4.1   G Study   72
              3.3.4.2   D Study   73
3.4   Summary   73

4   Factor Analytic Approach in Measurement   **75**
4.1   Chapter Overview   75
4.2   Introduction   75
4.3   Exploratory Factor Analysis (EFA)   76
      4.3.1   EFA of a Cognitive Inventory   77
      4.3.2   EFA Using the **psych** Package   89
      4.3.3   EFA with Categorical Data   91
4.4   Confirmatory Factor Analysis (CFA)   93
      4.4.1   CFA of the WISC-R Data   93
      4.4.2   CFA with Categorical Data 103
              4.4.2.1   Ordinal CFA–Method 1 103
              4.4.2.2   Ordinal CFA–Method 2 105
4.5   Summary 105

5   Item Response Theory for Dichotomous Items   **107**
5.1   Chapter Overview 107
5.2   Introduction 107
      5.2.1   Comparison to Classical Test Theory 107
      5.2.2   Basic Concepts in IRT 108
      5.2.3   IRT Model Assumptions 112
5.3   The Unidimensional IRT Models for Dichotomous Items . . . 113
      5.3.1   One-Parameter Logistic Model and Rasch Model . . . 113
              5.3.1.1   One-Parameter Logistic Model 113
              5.3.1.2   Rasch Model 119
      5.3.2   Two-Parameter Logistic Model 122

5.3.3 Three-Parameter Logistic Model 124
5.3.4 Four-Parameter Logistic Model 126
5.4 Ability Estimation in IRT Models 128
5.5 Model Diagnostics 133
5.5.1 Item Fit 134
5.5.2 Person Fit 136
5.5.3 Model Selection 139
5.6 Summary 141

6 Item Response Theory for Polytomous Items **143**
6.1 Chapter Overview 143
6.2 Polytomous Rasch Models for Ordinal Items 144
6.2.1 Partial Credit Model 144
6.2.2 Rating Scale Model 148
6.3 Polytomous Non-Rasch Models for Ordinal Items 151
6.3.1 Generalized Partial Credit Model 152
6.3.2 Graded Response Model 154
6.4 Polytomous IRT Models for Nominal Items 157
6.4.1 Nominal Response Model 158
6.4.2 Nested Logit Model 161
6.5 Model Selection 166
6.6 Summary 167

7 Multidimensional Item Response Theory **169**
7.1 Chapter Overview 169
7.2 Multidimensional Item Response Modeling 170
7.2.1 Compensatory and Noncompensatory MIRT 170
7.2.2 Between-Item and Within-Item Multidimensionality . 172
7.2.3 Exploratory and Confirmatory MIRT Analysis 174
7.3 Common MIRT Models 175
7.3.1 Multidimensional 2PL Model 175
7.3.2 Multidimensional Rasch Model 184
7.3.3 Multidimensional Graded Response Model 187
7.3.4 Bi-Factor IRT Model 189
7.4 Summary 192

8 Explanatory Item Response Theory **193**
8.1 Chapter Overview 193
8.2 Explanatory Item Response Modeling 193
8.2.1 Data Structure 194
8.2.2 Rasch Model as a GLMM 196
8.2.3 Linear Logistic Test Model 199
8.2.4 Latent Regression Rasch Model 203
8.2.5 Interaction Models 206
8.3 Summary 210

9 Visualizing Data and Measurement Models   **211**

9.1 Chapter Overview 211

9.2 Introduction 211

9.3 Diagnostic Plots 212

9.4 Path Diagrams 222

9.5 Interactive Plots with **shiny** 224

     9.5.1 Example 1: Diagnostic Plot for Factor Analysis 225

     9.5.2 Example 2: The 3PL IRT Model 228

9.6 Summary 231

10 Equating   **233**

10.1 Overview 233

10.2 Introduction 233

     10.2.1 Equating Designs 234

     10.2.2 Equating Functions and Methods 235

     10.2.3 Evaluating the Results 236

     10.2.4 Further Reading 236

10.3 Examples 236

     10.3.1 Equivalent Groups 237

         10.3.1.1 Identity, Mean, and Linear Functions 237

         10.3.1.2 Nonlinear Functions 241

     10.3.2 Nonequivalent Groups 243

         10.3.2.1 Linear Tucker Equating 243

10.4 Summary 247

11 Measurement Invariance and Differential Item Functioning   **249**

11.1 Chapter Overview 249

11.2 Measurement Invariance 249

     11.2.1 Assessing Measurement Invariance 250

         11.2.1.1 Configural Invariance 251

         11.2.1.2 Weak Invariance 251

         11.2.1.3 Strong Invariance 253

         11.2.1.4 Strict Invariance 254

         11.2.1.5 Assessing Partial Invariance 256

11.3 Differential Item Functioning 258

     11.3.1 The Mantel-Haenszel (MH) Method 259

     11.3.2 Logistic Regression 264

     11.3.3 Item Response Theory Likelihood Ratio Test 269

11.4 Summary 274

12 More Advanced Topics in Measurement   **277**

12.1 Chapter Overview 277

12.2 CRAN Task Views 277

12.3 Computerized Adaptive Testing 278

12.4 Cognitive Diagnostic Modeling 280

12.5 IRT Linking Procedures 280
12.6 Bayesian Models of Measurement 281
12.7 Hierarchical Linear Models 283
12.8 Profile Analysis 284
12.9 Summary 285

References **287**

Index **299**

Preface

Why This Handbook?

The fields of educational measurement and psychometrics focus on developing, validating, and analyzing assessments, tests, surveys, scales, and, more generally, instruments. Currently there are many great textbooks that provide comprehensive introductions to educational measurement and psychometrics, such as Brennan (2006), Thorndike and Thorndike-Christ (2010), and Lane, Raymond, and Haladyna (2015). Then what really made us write *Handbook of Educational Measurement and Psychometrics Using R*?

When we were graduate students in a quantitative methods program, running the exercises and activities in our measurement textbooks required many different software programs, such as WINSTEPS (Linacre, 2015), BILOG-MG (Zimowski, Muraki, Mislevy, & Bock, 2002), and PARSCALE (Muraki & Bock, 1997), many of which were commercial and too expensive for most graduate students. This situation was primarily due to the lack of a single software program capable of running the most widely used psychometric analyses (e.g., item and scale analysis, item calibration and scoring in item response theory, test equating, and differential item functioning).

During our graduate studies, the R programming language (R Core Team, 2017) really began to pick up steam in educational measurement and psychometrics. It was not very surprising that a growing number of researchers and practitioners in our field immediately started learning R. First and foremost, R is free, open-source, and cross-platform (e.g., available on Windows, Mac OS, and Linux). The built-in functions in R help us organize and manipulate data, run various statistical analyses, and create high-quality graphics. Although R is particularly useful for statistical computing and data visualization, it can also be used in a myriad of ways (e.g., data science and web programming). In addition, R is being actively developed and is growing exponentially, with a very attentive R core team and developers who submit packages to the Comprehensive R Archive Network (CRAN) and regularly update their algorithms in R, fix bugs, and add new enhancements.

In educational measurement and psychometrics, researchers and practitioners without computer programming skills often find R relatively easy to use. Therefore, many R users have begun to develop packages, specifically focusing on methods in educational measurement and psychometrics. A vast library of user-contributed packages enables everyone around the world to access a wide variety of statistical methods and data visualization tools. Currently, there are over 10,000 user-contributed packages in the CRAN database, available to researchers and practitioners from education, psychology, and other social science fields. The growing popularity of R has also motivated many authors to write tutorials, manuscripts, and textbooks focusing on measurement using R. For example, there are several R-related books on specific topics, such as latent variable modeling (e.g., Beaujean (2014) and Finch and French (2015)), item response theory (e.g., Baker and Kim (2017)), and survey data analysis (e.g., Falissard (2012)). However, there is no single book on R that covers a wide breadth of topics relevant to measurement and psychometrics.

In *Handbook of Educational Measurement and Psychometrics Using R*, we cover a variety of topics, including classical test theory; generalizability theory; the factor analytic approach in measurement; unidimensional, multidimensional, and explanatory item response modeling; test equating; visualizing measurement models; measurement invariance; and differential item functioning. Our book fills the niche of a general, all-purpose handbook of introductory and advanced measurement models using R. We focus on the widely used methods that our readers are likely to use in their research and work, not necessarily the latest and greatest methods in measurement.[1]

In conclusion, does this handbook focus solely on either R or educational measurement and psychometrics? Our answer would be no. With this handbook, we wish to help our readers learn enough R to be able to conduct psychometric analyses and build a good, solid foundation to further develop their R skills, and to refresh the memory of our readers with regard to the methods in measurement and psychometrics that may be relevant to their work. Using the analogy of a Swiss army knife, we would describe this handbook as a collection of data analytic tools in R that can guide researchers and practitioners when conducting psychometric analyses.

Who Is This Handbook Intended For?

This handbook is mainly intended for advanced undergraduate and graduate students, researchers, and practitioners, as a complementary book to an introductory or advanced textbook in measurement. We assume that either

---

[1]We talk about R packages for some of these methods in Chapter 12.

our readers have already had some training in educational measurement or psychometrics, or that they are concurrently taking a course on educational measurement or psychometrics. Practitioners and researchers who are familiar with the measurement models, but intend to refresh their memory and learn how to apply the measurement models in R, would find this handbook quite fulfilling. Students presently taking a course on educational measurement and psychometrics will find this handbook helpful in applying the methods they are learning in class. In addition, instructors teaching a course on educational measurement and psychometrics will find our handbook as a useful supplement for their course.

In this handbook, we have taken great care to ensure the validity and accuracy of our explanations regarding the measurement theories and techniques. However, as we have explained above, this handbook is not meant to cover the measurement topics in detail like a typical measurement textbook. In places where our readers may find our theoretical explanations too short, we recommend that our readers refer to either a general textbook on educational measurement and psychometrics or a textbook focusing on the topic of interest. There are lots of excellent textbooks on the methods and techniques in educational measurement and psychometrics; this handbook is just not one of them. Rather than being a primer on educational measurement and psychometrics, our handbook is intended to be an excellent resource on how to run and implement these methods in R, quickly and, hopefully, in a painless way.

Our handbook begins with the basics of R. We assume that our readers have no experience with R, have never even downloaded it, and have only heard of it in passing. We hope that our examples and code snippets will be useful for overcoming any command-line anxiety that R may induce, and that our readers will have an arsenal of relevant code that they can readily edit and adapt for their own practice. Readers who are already advanced R users and are referencing this book for measurement will surely learn something new, but the pace of the examples (i.e., the detailed code explanations) may be a bit slow for them. Readers who are already accustomed to R and know how to employ the psychometric methods in R might find this handbook too basic. Although we appreciate any support from such readers, and think that they might also pick up some tricks presented here and there, it is likely that they already know how to run many of the methods presented in this handbook.

Finally, some knowledge of and comfort with statistics and statistical notation is helpful for understanding this handbook. However, readers with weaker mathematical and/or statistical backgrounds may feel free to skip the sections on theory in the chapters and move directly to the illustrated examples in R. Regardless, we strongly recommend that all of our readers run the R code presented in this handbook by following the examples, and adapt the code to their data and research needs.

Layout of Our Handbook

Learning R from scratch can be slow and cumbersome for readers who have no programming experience or are more familiar working with point-and-click statistical software programs. Therefore, we strongly recommend that our readers who have never used R before or are novice R users review Chapter 1, where we explain how to install R, how to install and activate R packages, and how to execute R code in detail. In addition, we demonstrate the use of some basic R functions for entering, manipulating, and summarizing data. We believe that Chapter 1 will help our readers build a good foundation in R and better digest the content in the rest of this handbook.

Each chapter begins with a general overview of a particular measurement topic, followed by the applications of that topic using R. Throughout the handbook, we aim to create a balance between the text, statistical notation, and R code. In each chapter, we review measurement theories and psychometric techniques, but do not introduce them in an overly detailed way. We provide a brief, but not exhaustive, review of a method and then explain the implementation of the method using the R code in plain English. We hope that by reading this handbook, our readers will learn enough about R to adapt our examples for their data sets and research problems regarding educational measurement and psychometrics.

Notation

Throughout the handbook, we use **bolded** fonts to refer to R packages. For example, the R package **lme4** will be bolded throughout the chapters. We do this to draw the readers' attention to the name of the package. When we are referring to R code, we use the `typewriter` font. When we are referring to names of variables within a data frame or matrix, we will use this normal font, unless we write `data$variable.name`. Therefore, we will reserve `typewriter` font for valid R code only.

As we will mention again in Chapter 1, we present R code as chunks. The code that we want our readers to type in R (i.e., the input chunks) will look like this:

```
2 + 2
```

Unlike the input chunks, the chunks that contain the output (i.e., the output chunks) will be grayed like this:

4

Our readers should only type the code in the input chunks into R, not the code in the output chunks.

Acknowledgments

Chris would like to thank the Center for Applied Research and Educational Improvement at the University of Minnesota for supporting his work on this handbook.

Okan would like to thank his students at the University of Alberta for showing a high interest in learning R and motivating him to continue teaching educational measurement and psychometrics using R.

We both would like to thank our professors in Quantitative Methods in Education, Quantitative Psychology, and Statistics at the University of Minnesota for inspiring us to learn the R programming language and expand our knowledge of educational measurement and psychometrics.

We both would like to thank CRC Press for their patience and steadfast support, as this was the first book we have authored and we were sometimes a little too optimistic when it came to deadlines.

Finally, we both would like to thank the R Core Team and the R community for writing and maintaining exceptional, high quality packages. The innovative and prolific R ecosystem makes writing a book like this plausible, a pleasure, and also quite a challenge!

List of Figures

1.1   Default RStudio setup after creating a new R script.   5
1.2   Default diagnostic plots from a multiple regression model. . .   25
1.3   Scatter plot matrix of verbal measures in the interest data set   26
1.4   Scatter plot of vocab by reading conditional on gender   28

2.1   Dot plot of age by sex of examinees taking the interest inventory   39
2.2   Q–Q plot of the vocab variable in the interest data set   40
2.3   Empirical distribution for coefficient alpha (n = 10,000) . . .   46

3.1   Venn diagram of a two-facet crossed (p x w x r) design. Note that s stands for student (the unit of measurement), w stands for writing prompt, r stands for rater, and e stands for error.   59
3.2   Venn diagram of a one-facet (p x i) design. Note that p stands for participant (the unit of measurement), i stands for item, and e stands for error.   61
3.3   Histogram of the percent of items correct on the EF instrument   63
3.4   Plot of dependability coefficient against number of EF items for the one-facet random design   66
3.5   Plot of generalizability coefficient against number of writing prompts by number of raters   69
3.6   Venn diagram of a two-facet, partially nested design, (r:s) x w. Note that s stands for student (the unit of measurement), w stands for writing prompt, r stands for rater, and e stands for error.   70

4.1   Histograms of indicators of cognition   80
4.2   Scatterplot matrix of the cognition measures   81
4.3   Bollen plot of cognition data   82
4.4   Scree plot of the cognition dataset   85
4.5   Parallel analysis of the cognition dataset   86

5.1   Item characteristic curves of the items with easy, moderate, and high difficulty.   109
5.2   Item characteristic curves of the items with low, moderate, and high discrimination.   110
5.3   Item characteristic curves of the items with low and high guessing.   111

5.4   The reciprocal relationship between TIF (the solid line) and cSEM (the dashed line) 112

5.5   Item characteristic curves for items 1 (reason.4) and 2 (reason.16) for the 1PL model 117

5.6   Item information functions for items 3 (reason.17) and 5 (letter.7) for the 1PL model 118

5.7   Combined ICCs for items 1 (reason.4) and 2 (reason.16) for the 1PL model 119

5.8   The TIF and cSEM plot for the 1PL model 120

5.9   Item characteristic curves for items 2 (reason.16) and 16 (rotate.8) for the Rasch model 122

5.10   Item characteristic curves for items 12 (matrix.55) and 14 (rotate.4) for the 2PL model 124

5.11   Item characteristic curves for items 5 (letter.7) and 16 (rotate.8) for the 2PL model 125

5.12   Item characteristic curves for reason.4 (item 1) and reason.19 (item 4) for the 3PL model 127

5.13   Scatterplot matrix of latent trait estimates 132

5.14   Empirical plot for item 1 137

5.15   Distribution of the Zh statistic for the Rasch model 139

6.1   Thresholds between the four ordered response categories. . . 145

6.2   Option characteristic curves for items 2 and 5 for PCM . . . 148

6.3   Item information functions for items 2 and 5 for PCM 149

6.4   Location and threshold parameters of two rating scale items. 150

6.5   Option characteristic curves for items 2 and 9 for RSM 152

6.6   Option characteristic curves for items 6 and 8 for GCPM . . 154

6.7   Cumulative thresholds between the four response categories. . 155

6.8   Option characteristic curves for Q5 and Q9 for GRM 157

6.9   Option characteristic curves for items 3 and 15 for NRM . . . 161

6.10   Option characteristic curves for item 15 for NRM 162

6.11   Option characteristic curves for items 8 and 21 for 2PL-NLM 164

6.12   Option characteristic curves for items 1 and 17 for 3PL-NLM 166

7.1   Item characteristic surface for a compensatory MIRT model. 172

7.2   Item characteristic surface for a noncompensatory MIRT model. 173

7.3   Between-item (left) and within-item (right) structure. 174

7.4   Item surface plot for item13 for the M2PL model. 180

7.5   Item contour plot for item13 for the M2PL model 181

7.6   Item information function plot for item13 for the M2PL model. 182

7.7   Test information function (left) and cSEM (right) plots for the M2PL. 182

7.8   Item surface plots for item7 (left) and item13 (right) for the MGRM. 189

7.9   Bi-factor test structure.   190

8.1   Item-person map for the Rasch model   200

9.1   Mahalanobis distance plotted against the observation identifier   214
9.2   Generalized Cook's distance plotted against observation ID .   216
9.3   Normal Q–Q plot comparing standardized residuals for the six manifest variables against normal theoretical quantiles   219
9.4   Manifest variables plotted against the estimated factor scores   220
9.5   Standardized residuals against ids by the six manifest variables   221
9.6   Path diagram for the two-factor cognitive model   223
9.7   Path diagram of the multi-group, configural model for the two-factor cognitive model   224
9.8   Screenshot of the **shiny** application for the factor analysis diagnostic plot.   228
9.9   Screenshot of the **shiny** application for the IRT plot.   231

10.1 Bar plot of the test scores on form X in the hcre data set . .   239
10.2 Scatterplot of the adjusted X scores on form Y against the original form X scores   242
10.3 Plot of the common anchor scores against total scores on form X   244
10.4 Smoothed (loglinear) plot of the common anchor scores against total scores on form X   245

11.1 Uniform (left) and nonuniform (right) DIF in a dichotomous item.   259
11.2 Flagged items based on the MH chi-square test.   265
11.3 Flagged items based on the logistic regression approach. . . .   270
11.4 The item characteristic curves for item 6 (S2WantShout) in the VerbAgg data set.   271
11.5 Item characteristic curves showing uniform DIF in items 14 and 16   274

List of Tables

1.1 Commonly used functions in R and in this handbook. 10

1.2 Packages used in this handbook. 13

9.1 Standardized parameter estimates from the cognitive model using the full `interest` sample and without observation 111. . 217

10.1 Available equating functions for each equating method (adapted
from Albano (2016)). 243

11.1 A $2 \times 3 \times 2$ contingency table for the MH method. 260

1

Introduction to the R Programming Language

1.1 Chapter Overview

In this chapter, we provide a brief introduction to the R programming language that will be used throughout *Handbook of Educational Measurement and Psychometrics Using R*. We first introduce R and show our readers how to download and install both R and RStudio. The remainder of the chapter walks through running R code in RStudio and introduces the **hemp** package, the companion package for this handbook. The **hemp** package contains the data sets and many of the functions that are used in this handbook. The package also automatically installs some of the R packages necessary for running the code used in this book. Our readers can find a complete list of the R packages used in this handbook under the "R Packages" section in this chapter.

1.2 What Is R?

The R Project for Statistical Computing (https://www.r-project.org) supports and develops R, a free and open-source language and environment for statistical computing and graphics. R is cross-platform and available on Windows, Mac OS X, and various UNIX-type platforms (e.g., Linux, FreeBSD, and OpenBSD). It is the premier statistical language and the statistical environment that will be used in this handbook.

There are a multitude of reasons why our readers may be interested in learning R:

1. R is free and open-source. Because R is free, everyone can have access to it provided they have a computer and an internet connection. Because R is open-source, statisticians and computer scientists can

completely audit the codebase to remove any errors (e.g., bugs and security flaws) and to increase computing performance. This means that the code in the core of R can really be trusted. Absolutely nothing in R has to be a black box.

2. R is the computing environment of statistics. When one reads about a new method in a statistics, data mining, machine learning, or psychometric article, the odds are someone has already written a package or code that implements the algorithm or method in R. It is not unusual in the statistical world that when the authors submit a new method to a statistics journal, an R package accompanies that submission. There is no need to learn specialized software. By just learning R well, our readers will be able to perform most analyses.

3. R has amazing visualization capabilities. This includes the R base graphics as well as the **lattice** and **ggplot2** packages.

4. The R community. There are some extremely talented and prolific programmers in the R world that have contributed a tremendous number of packages to R. In addition, many of these programmers and users have taken the time to provide support either on the R mailing list or Stack Overflow (https://stackoverflow.com/) where most R users seek further help or clarification regarding their R code.

5. R has a great integrated development environment, RStudio. In addition to supporting R, RStudio also supports Python, C/C++, LaTeX, Markdown, JavaScript, and more. RStudio makes learning R easier and helps users gain better programming and coding habits.

Learning R can be difficult and frustrating at first, especially for those readers who do not use it regularly or have no programming experience. However, we can assure our readers that the hard work put into learning R will allow them to organize and manage data more efficiently, run a variety of statistical models, create beautiful, high-quality plots, and write reports quicker than they could ever do in a statistical software program with a graphical user interface (GUI). In addition, we strongly believe that R's syntax helps reinforce the concepts and methods in measurement and psychometrics as it often mimics the mathematical notation used in these fields.

1.2.1   Our Approach to R

In general, we take a conservative approach to teaching R. That is, in this handbook we generally rely on the R base packages. These are the default packages that come with a basic R installation. Our reason for this approach is that the functions provided by the R core team are the most thoroughly vetted functions and that the R core team tends to be conservative in their development, meaning that R code written now will likely work in 10 years without any modification. Unfortunately, we cannot always rely on core packages for all the models presented here. In those situations, we select packages that are commonly used in our field, are feature rich, and are currently being maintained and developed.

The negative side of this approach to teaching R is that many of the functions that make R easier to use, learn, and read (e.g., many of the packages in the **tidyverse**[1] and the **magrittr** package) will not be discussed in this handbook. However, what our readers learn from this handbook will be of great value to them regardless of whether they decide to stick with the core functions, move on to more specialized packages for data management, or write their own functions. For our readers who might interested in an alternative approach to learning R, we recommend Wickham and Grolemund (2017).

1.3 Obtaining and Installing R

To obtain R, the reader must visit the Comprehensive R Archive Network (CRAN) website,

https://cran.r-project.org/

or search for CRAN in in their web browser. The CRAN website provides pre-compiled binaries of the base R system and contributed packages for Windows, Mac, and Linux. The CRAN website selects a mirror (a server) automatically where the user is able to download R. If our readers would like to manually select a mirror that is geographically closer, then one of the mirrors from the CRAN list can be specified.[2] For example, if someone lives in Boston, then either the Pittsburgh or Ohio mirrors might be a sensible option. Selecting a mirror that is geographically closer can increase the speed of the download. Our only recommendation to the reader is to select an HTTPS link over an HTTP one.[3]

Once a mirror has been selected, a new page will open that allows R to be downloaded. Below, we provide platform-specific installation instructions for downloading and installing R.

---

[1] http://tidyverse.org/
[2] https://cran.r-project.org/mirrors.html
[3] See https://goo.gl/B1hLEw for more details.

1.3.1   Windows

To download R for Windows, Windows users should click "Download R for Windows." This will load the R for Windows page where the user should click the "base" link. This will then load an R version specific website (e.g., R-3.4.3 for Windows (32/64 bit)) where the user should then click on "Download R 3.4.3 for Windows (62 megabytes, 32/64 bit)" (or whatever is the current version). This will download the R installer for Windows. Once the R installer is downloaded, the user double-clicks the installer. If R has been installed correctly, clicking the R shortcut in the start menu should open the R console.

1.3.2   Mac

To download R for Mac OS X, Mac users should click "Download R for (Mac) OS X". This will load the R for Mac OS X page. Under the "Files:" header, Mac users should click the link for the first pkg file as this will be the most recent R release. If R 3.4.3 is the most recent version, this link will be called "R-3.4.3.pkg." Clicking this link will download the installer. Once the download is complete, the user should double-click the installer to install R. Once R is installed, it can be launched by locating R in Applications. If R has been installed correctly, then the R console should open.

1.3.3   Linux

To download and install R for Linux, Linux users should click "Download R for Linux." The users will then be redirected to a page with links for installation instructions for Debian, Redhat, SUSE, and Ubuntu. For other Linux distributions, we recommend doing a search of the name of their distribution and R. Many Linux distributions package R, but are not listed on CRAN. An R installation on Linux, and other UNIX-like operating systems, does not include an editor, unlike the Windows and Mac versions, and consists of just the console-based R program. R can be launched from a terminal by typing R in a terminal.

1.4   Obtaining and Installing RStudio

RStudio is an integrated development environment for R and we strongly recommend that both new and experienced R users consider RStudio instead of working directly with the R program provided by the R project. There are

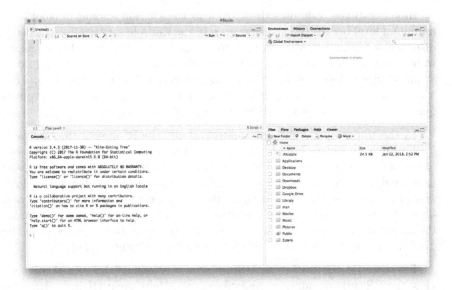

FIGURE 1.1
Default RStudio setup after creating a new R script.

a myriad of benefits of using RStudio over working directly with R's GUI, but the most important benefit among these is that it provides a uniform interface and experience across all the operating systems.

RStudio is free and open source, but there is also a paid, commercial version. We have no experience with the latter and the free, open source version is the one that is used throughout this handbook. To install RStudio, our readers should open the following link in a web browser:

https://www.rstudio.com/products/rstudio/download/

and then scroll to the bottom of that page until they see the "Installers for Supported Platforms" header and select the appropriate installer based on their operating system. Once RStudio is installed, it should be launched, and a new R script should be created from the File menu on the top left corner (File → New File → R script). What our readers would see should be similar to Figure 1.1.

RStudio is divided into four panes by default. The upper-left pane is the Source pane, which contains the script editor. In this pane, the R code is typed, and is then sent to the Console (the lower-left pane) by clicking "Run" or by using a keyboard shortcut (see Help → Keyboard Shortcuts Help). There is also a tab in the lower-left pane that provides easy access to the Terminal, which new R users would not need.

In general, the code directly written in the Console should be kept at a minimum level because any code written in the Console is typically considered dispensable (i.e., it would never need to be run again). The code that we might want to run again in the future should be written in the Source pane and saved as an Rscript before closing RStudio (as one would save any other file). That way, whenever RStudio is opened and the script is re-run, the same output is quickly and easily recreated, without typing all the code again.

In the upper-right pane, information about the R objects that we have in the Environment (more on this later) can be found as well as the History of the code that we have run in the Console. Finally, the bottom-right pane lists the files in our current working directory, displays plots, allows us to load and install packages without needing to run any code, view help pages, and interact with JavaScript-based visualization. The reader can easily re-organize the layout of these panes (e.g., switching the positions of the panes from the Tools menu on the top; Tools → Global Options → Pane Layout).

For more information about RStudio, we recommend that our readers review the RStudio documents available via the Help menu (Help → RStudio docs) as well as webinars, cheatsheets, and other learning materials available on the RStudio website (https://www.rstudio.com/).

1.5   Using R

This introduction is not meant to be an exhaustive tutorial about how to use R. However, we want to get our readers started using R (assuming they have never used R before). We introduce our readers to some common ways that R can be used. In the following chapters, we continue to introduce and discuss different applications using R.

We would like to note that the R programming language is quite flexible, and thus there are so many different ways of completing a given task in R. This also implies that there are so many functions (either built-in or user defined) and arguments necessary for using these functions. While we wish that our readers (especially those who are new to R) run all the code presented in this handbook, we do not expect them to remember all the details about the functions and understand immediately what all the code actually does.

Just like learning a foreign language or a musical instrument, learning the R language also takes a while. Therefore, we hope that as our readers follow the examples presented throughout this handbook, they focus on the whole picture rather than getting stuck on the details. Being patient definitely pays off while learning R.

Unless explicitly stated, we assume that all R code is written directly into the Source pane and run via the Run button, the Code menu, or a keyboard shortcut, and not entered directly into the Console. This will allow the reader to build a script of R code that they can reuse and modify for their own projects, whenever they need to do so.

1.5.1 Basic R Usage

The R code in this handbook is written in code chunks. Input chunks that we want the reader to type into the Source pane will look like this:

```
2 + 2
```

while output chunks that contain the output will look like the following:

```
[1] 4
```

Notice that the input chunks have dark shading, while the output chunks are lighter. The readers should **NOT** type the output chunks into the Source panes.

In R, comments about what the code is doing, or the purpose of the R script or function, are specified with a hashtag (#). When running multiple lines of codes, any line starting with # is considered a comment in R, and is not executed. This is a useful feature because it allows adding annotations either before or after the code, which can help us remember the purpose of the code when we use it again in the distant future.

In the following example, R ignores the first line because it is a comment, but executes the second line.

```
# The following R code calculates the sum of 2 and 2
2 + 2
```

If the readers either type or copy and paste the above chunk into the Source pane, highlight both lines of code, and click "Run," they will see the following result in the Console.

```
> # The following R code calculates the sum of 2 and 2
> 2 + 2
[1] 4
>
```

The Console echoes the input that has been sent from the Source pane. The input in the Console, but not in this handbook, is preceded by a >, the command prompt. We have removed this from our output chunks, as well as the input commands, to make the output easier to read. The final point to note is that sometimes the output in the Console is preceded by a [1], as shown in the example above.

R can be used like a calculator to do addition, subtraction, multiplication, and division. For example,

```
6 + 20
89 - 53
424 * 68
1110 / 37
```

In addition, R provides many built-in mathematical functions, such as logarithms. The following code calculates the natural log of 10.

```
log(10)
```

To save the output of a function to a variable in R, we use the assignment operator, <-. To create a variable x that equals 1, we need to type

```
x <- 1
```

If we want to print x, we just type and run x, and R will return 1.

```
x
```

```
[1] 1
```

Variables in R are typically called objects. Objects may contain a single value; a collection of values, called a vector; an entire data set, called a data frame; the output from a statistical model; a graphical plot; and so on. Nearly everything in R can be saved using the assignment operator and everything in R is an object. In this handbook, we sometimes refer to saving model output to an object. This is equivalent to saying that we saved 1 as the object x in the above chunk.

An important feature of R that our readers must certainly be aware of is that R is case-sensitive. So, R code written in lowercase would not refer to the same code written in uppercase. Continuing from our previous example where we have defined x as 1, we see if X would return the same result as x. When we type X and run the code as

```
X
```

```
Error: object 'X' not found
```

R produces an error message, indicating that this object does not exist. But, typing x and running it in R as

```
x
```

```
[1] 1
```

prints the number 1 as expected.

To turn x into a vector containing multiple values, we use the c function, which stands for combine,

```
x <- c(1, 5, 3, 6, 9)
```

This saves values 1, 5, 3, 6, and 9 to the x object. This operation also replaces the previously defined x, which only included the number 1, with a new object consisting of a set of values. This is an example of how we can overwrite our existing objects in R. Sometimes this may cause errors in our code as we may forget what has been assigned to the object (e.g., is x a scalar that is equal to 1 or a vector containing 1, 5, 3, 6, and 9?). We strongly recommend against overwriting existing objects. Instead of overwriting, our readers should choose a unique name for each object. Typical conventions for naming objects in R include:

1. all letters in lowercase (i.e., `mydata`),

2. all letters in uppercase (i.e., `MYDATA`),

3. using camelCase (i.e., `myData`),

4. a combination of letters and numbers[4] (i.e., `mydata10`),

5. a period to separate words and numbers[5] (i.e., `mydata.10`), and

6. an underscore to separate words and numbers[6] (i.e., `my_data`).

We recommend that our readers choose a particular naming convention and use it consistently throughout their R scripts, to minimize any errors due to naming objects.

In the above example, `x <- c(1, 5, 3, 6, 9)` was defined as a numeric vector. Vectors containing strings of characters can also be saved as an object in R. For example, if we wanted to create a variable called **gender**, which contains information about participants' gender in a study, we would do the following:

```
gender <- c("male", "male", "male", "female", "female")
```

By default, R will save this information as a character vector. Sometimes we need these numeric and character vectors to be saved as factors for the psychometric models and analyses that we perform in R. We can convert numeric and character vectors to a factor by:

```
x_f <- as.factor(x)
gender_f <- as.factor(gender)
```

---

[4]An object name cannot start with a number in R (e.g., `5mydata` would not be an acceptable name).

[5]An object name cannot start with a period in R (e.g., `.mydata` would not be an acceptable name).

[6]An object name cannot start with an underscore in R (e.g., `_mydata` would not be an acceptable name).

TABLE 1.1
Commonly used functions in R and in this handbook.

Function	Meaning
abs(x)	the absolute value of x
exp(x)	exponential of x
log10(x)	log of x using base 10
sqrt(x)	square root of x
x^y	raises x to the power of y (e.g., $x\hat{2}$)
cor(x, y)	correlation of x and y
cor(x)	correlation matrix of x
var(x)	covariance matrix of x
sd(x)	standard deviation of x
mean(x)	mean of x
median(x)	median of x
sum(x)	sum of the values in x
max(x)	maximum value in x
min(x)	minimum value in x
sort(x)	sort the values in x in ascending order
apply(x, 1, function)	performs the function for each row in x
apply(x, 2, function)	performs the function for each column in x
ifelse(x == y, yes, no)	if x is equal to y, do yes, else do no

It is also possible to combine multiple objects together. For example, two objects (i.e., y and z) can be combined as a new object (i.e., yz) as:

```
y <- c(9, 3, 4)
z <- c(2, 6, 3)
yz <- c(y, z)
```

Objects that perform operations in R are referred to as functions. Functions take an argument and return one or more values. The typical syntax for using an R function is:

```
function_name(argument1, argument2, ...)
```

where function_name could be c, as.factor, log, or so on; argument1 and argument2 refer to the first and second arguments; and ... refers to additional arguments. Some of the R functions that are often used in statistical and psychometric analyses are listed in Table 1.1.

In the functions we used so far, we have passed only a single argument; however, many functions take more than one argument. For example, the log function will allow us to compute a base 10 logarithm if we pass base = 10 as a second argument.

```
log(10, base = 10)
```

For most built-in and user-defined functions, the authors include examples to demonstrate how their functions can be used. These examples can be seen using the `example` function, as illustrated below.

```
example(log)
```

1.5.2   R Packages

A suite of related R functions are often combined into an R package or library. A default R installation comes with a set of packages and functions that allow us to run various statistical models, create visualizations, and organize data. However, we may need or want to install an additional R package to perform a specialized analysis (e.g., confirmatory factor analysis or item response theory), to use a more advanced, customizable graphical visualization framework, or to more efficiently manage and organize our data. Many of these supplemental packages are authored by various statistical and computational experts throughout the world and will greatly enhance the R experience and improve on computational speed.

There are three common repositories where authors of R packages can host their packages: CRAN, GitHub, and Bioconductor.[7] To install packages that are hosted on CRAN, where most R packages can be found, we use the `install.packages` function. For example, to install the item response theory package **mirt**, which is used in Chapters 5 through 7, we would run the following command:

```
install.packages("mirt")
```

Once we have installed the package, we do not need to run the `install.packages` function again unless we remove the package or install a newer version of R. From time to time, authors of R packages add new functions or update their existing functions to fix bugs. Once a package is updated on CRAN, we can obtain the latest updates for this package using the `update.packages` function (assuming we already had an older version of the package installed on our computer).

```
update.packages("mirt")
```

Once a package is installed, we need to activate the package so that the functions in the package are available to be used in the current R session.

---

[7]We will not use Bioconductor in this handbook as its functions are geared towards bioinformatics and genomic data. More information about Bioconductor is available at https://www.bioconductor.org/install/

Continuing from the example above, we would activate the **mirt** package as shown below:

```
library("mirt")
```

To find out more information about the **mirt** package, we can type mirt into the search bar (where the magnifying glass icon is located) in the lower-right pane, or type the following in the R console:

```
library(help = mirt)
```

To see a complete list of the functions in the **mirt** package:

```
ls("package:mirt")
```

and to see the data sets included with the **mirt** package:

```
data(package = "mirt")
```

GitHub (https://github.com) is a software development platform. Many R developers use GitHub to develop and host their packages temporarily, which they later submit to CRAN. Also, some developers do not bother with CRAN and permanently host their packages on GitHub. This is because CRAN has a strict policy for creating and maintaining R packages, while GitHub does not. Generally, we can consider packages on CRAN to be stable versions, while packages on GitHub may be in active development and in beta.

To install a package from GitHub, we first need to install and activate the **devtools** package (Wickham & Chang, 2017).[8]

```
install.packages("devtools")
library("devtools")
```

Next, we use the `install_github` function in the **devtools** package to be able to install any R package hosted on Github.

Throughout this handbook, we use the **hemp** package, which is currently available on GitHub. The **hemp** package is a companion package for this handbook and contains all the data sets as well as functions for performing analyses and creating plots used in this handbook. In addition, the **hemp** package acts as meta-package, installing and loading some of the other packages that are used in this handbook.

To install **hemp** and then activate it, the reader should run the following commands:

```
install_github("cddesja/hemp")
library("hemp")
```

---

[8]For some computers, the **curl** package might also be necessary. To install the **curl** package: `install.packages("curl")`

TABLE 1.2
Packages used in this handbook.

Package	Purpose	Version
boot	bootstrapping	1.3–20
difR	differential item functioning	4.7
equate	equating	2.0.6
faoutlier	detecting/visualizing influential points	0.7.3
GPArotation	rotations for factor analysis	2014.11–1
lattice	trellis graphics	0.20-38
lavaan	latent variable modeling	0.5–23.1097
lme4	general & generalized mixed-effect models	1.1–13
mirt	item response theory	1.25
psych	general psychometrics	1.75
semPlot	visualizing path diagrams	1.1
shiny	web application for R	1.0.5
hemp	general psychometrics	0.1.0

In addition to **hemp**, we use a series of packages throughout the chapters. Table 1.2 lists all the packages, and their version numbers, used in the examples presented in this handbook. All of these packages, with the exception of **hemp**, are available on CRAN.

Before running any of the examples or code in any of the chapters, the **hemp** package and any package mentioned at the beginning of the chapter must be loaded. If the **hemp** package, and the other required packages for the examples, are not installed and activated, then an error message will occur when running the examples in Chapters 2 through 11.

All the R code used in the examples presented throughout this handbook can be downloaded in a zipped folder at `http://bit.ly/hemp_code`.

1.5.2.1 Masked Functions

Sometimes two packages in R contain the same function name, and when they are both activated within the same R session, the package that has been activated second will overwrite, or mask, the function from the first package. For example, both **lme4** and **mirt** contain a function called `fixef`. We will see a message in the Console about this when the packages are initially activated, which is quite easy to miss. When both packages are activated, if we type:

```
?fixef
```

we will see that in the lower-right pane of RStudio (under the Help section), we are prompted to select the `fixef` function's help page that we are trying to reach. If we are trying to use **lme4**'s `fixef` function but have activated **mirt** second, we will get a cryptic message when running the `fixef` function:

```
Error: Only applicable to MixedClass and SingleGroupClass
objects
```

means that although we are trying to use the `fixef` function in **lme4**, R is using the `fixef` function from **mirt**. To avoid this issue, we can access a function directly from a package by using the double colon sign (::)

```
lme4::fixef(x)
```

This will allow us to call the `fixef` function in **lme4** as opposed to **mirt** or any other package we have activated after **lme4** that provides a `fixef` function.

Because some of the R packages in this handbook have the same function names, it is possible that if our readers activate the packages in an order different from ours, they may see an error message similar to the one above. If the reader is not able to replicate our findings and gets an error message, we recommend checking to see if it is because a function is being masked by another function in a different package. To check this, one should type `package_name::function_name` directly as we have demonstrated above for the `fixef` function in **lme4**.

1.5.3   Assessing and Reading in Data

Authors of R packages often include example data sets in their packages to demonstrate the functions available in the package. Once a package is installed, the example data sets are also installed along with the functions in the package. To able to use these data sets, activating the package is necessary. Generally, when a package is activated in R, its example data sets are also automatically activated.

For data sets that are not automatically loaded in R, we use the `data` function. For example, we can activate the `interest` data set (a cognitive, personality, and vocational interest inventory) from the **hemp** package as follows:

```
library(hemp)
data(interest)
```

If we wanted to activate an additional data set, e.g., the `HSQ` data set from **hemp**, we would type:

```
data(HSQ)
```

To see the currently active data sets in R, and any objects that we have created, we can run the `ls` function.

```
ls()
```

If we want to export and save a data set from R to a spreadsheet to be read in Excel (or a text editor), we can use the `write.csv` function. In the following example, we save the `interest` data set into a spreadsheet called "interest.csv." We specify the `row.names = FALSE` argument to prevent R from printing the row names in the `interest` data set as an additional variable.

```
write.csv(interest, file = "interest.csv", row.names = FALSE)
```

This procedure saves "interest.csv" in our working directory. To find out where the interest.csv file has been saved (i.e., the path to our working directory), we need to type

```
getwd()
```

We can change the location of our working directory through the RStudio menu (Session → Set Working Directory → Choose Directory). Setting the working directory before running code in R is generally a good idea because when we save our data, model output, plots, etc., we may not know where they are saved unless we explicitly state the file location.

Data in a spreadsheet (i.e., the .csv format) can be read back into R using the `read.csv` function. Below, we read the interest.csv file back in R, specifying that there is a header row (i.e., the first row in the file contains the variable names), that we want to treat any strings as character vectors and not as factors, and that we wish to save the object in R as `interest_new`.

```
interest_new <- read.csv("interest.csv", header = TRUE,
                         stringsAsFactors = FALSE)
```

If the interest.csv file is not located in the current working directory, then we need specify the path to the file so that R can find the file on the computer.

Assume that the interest.csv file is located in the Desktop folder (e.g., "C:\Users\User\Desktop\interest.csv") in a computer with a Windows operating system.[9] To see the full path for a file, readers who use a Windows operating system can right-click on the file and select "Properties"; readers who use a OS X can right-click on the file and select "Get Info." In the example below, we first replace the backslash in the file path (\) with a forward slash (/); otherwise, R would not be able to read the file path properly. Then, we specify the full file path in the `read.csv` function as:

---

[9]The file path would look like "/Users/User/Desktop/" in a computer with Mac OS X or Linux.

```
interest_new <- read.csv("C:/Users/User/Desktop/interest.csv",
                         header = TRUE,
                         stringsAsFactors = FALSE)
```

If there is an issue with the import process, then the R console typically shows an error message indicating that R has not been able to find the file or read the file properly. In addition to spreadsheets, SPSS and SAS XPORT files can also be read in using the **read.spss** and **read.xport** functions in the **foreign** package. More information about these functions can be found using the following commands:

```
library("foreign")
?read.spss
?read.xport
```

Because all of the data sets that we use in this handbook are already included in the **hemp** package, we will not spend more time on this topic. However, readers wishing to learn more about how to read external data into R may find the Cookbook for R[10] and the Stat Methods[11] websites helpful.

1.5.4   Data Manipulation

After data have been read or activated in R, it is important to examine the content and structure of the data. The **head** function can be used to view the first rows of the data. In the following example, we view the first three rows of the **rse** data set in the **hemp** package (assuming that the **hemp** package has already been activated using `library("hemp")`).

```
head(rse, 3)
```

	Q1	Q2	Q3	Q4	Q5	Q6	Q7	Q8	Q9	Q10	gender	age	source	country
1	2	2	3	3	2	2	3	3	3	3	1	10	1	IN
2	2	2	0	2	1	1	1	1	0	0	1	16	1	US
3	1	1	0	1	0	0	0	0	0	0	2	17	1	NL

	person
1	1
2	2
3	3

If the number of rows to be view is not specified (e.g., `head(rse)`), the **head** function returns the first six rows of the data set. We can also view the last six rows of the data with the **tail** function:

---

[10]http://www.cookbook-r.com/Data_input_and_output/
[11]https://www.statmethods.net/input/importingdata.html

```
tail(rse)
```

The `rse` data set is in a format known as a data frame in R. To get a complete list of the variable names in a data frame, we can use the **names** function.

```
names(rse)
```

```
 [1] "Q1"      "Q2"      "Q3"      "Q4"      "Q5"
 [6] "Q6"      "Q7"      "Q8"      "Q9"      "Q10"
[11] "gender"  "age"     "source"  "country" "person"
```

To access a particular variable within a data frame, we type the name of the data frame, a $ sign, and the name of the variable. For example, to access and print all the responses to the country variable, we would type:

```
rse$country
```

To determine the structure of the variables in the `rse` data set, we can use the `str` command.

```
str(rse)
```

```
'data.frame': 1000 obs. of  15 variables:
 $ Q1      : int  2 2 1 2 2 3 2 3 0 2 ...
 $ Q2      : int  2 2 1 2 2 3 2 3 3 2 ...
 $ Q3      : int  3 0 0 2 2 3 2 3 3 2 ...
 $ Q4      : int  3 2 1 2 2 3 2 3 3 2 ...
 $ Q5      : int  2 1 0 2 1 2 2 3 3 2 ...
 $ Q6      : int  2 1 0 1 1 2 1 2 3 2 ...
 $ Q7      : int  3 1 0 1 1 2 1 2 3 1 ...
 $ Q8      : int  3 1 0 1 1 3 1 2 3 1 ...
 $ Q9      : int  3 0 0 1 0 3 2 1 3 1 ...
 $ Q10     : int  3 0 0 1 1 3 2 2 3 1 ...
 $ gender  : int  1 1 2 1 1 2 2 1 1 2 ...
 $ age     : int  10 16 17 36 15 40 0 30 0 39 ...
 $ source  : int  1 1 1 3 3 3 1 1 3 1 ...
 $ country : Factor w/ 76 levels "A2","AE",..: 34 73 49 ...
 $ person  : int  1 2 3 4 5 6 7 8 9 10 ...
```

The output shows that all of the variables, except for country, are integers (labeled as "int" in the output), while country is a factor. Earlier in this chapter, we demonstrated how to convert a variable from a character variable to a factor. To convert a factor back to a character variable, we would use the `as.character` function. In the following example, we convert country into a character variable, which replaces the original country variable in the `rse` data set with this new character variable.

```
rse$country <- as.character(rse$country)
```

After performing this conversion, we can use either the `str` function or the `class` function to confirm the change. For example, we can confirm that `rse$country` is now a character vector:

```
str(rse$country)
class(rse$country)
```

To add a new variable into an existing data set, we would type:

```
rse$id <- 1:nrow(rse)
head(rse)
```

This adds a new variable, id, and assigns it the value of 1 through (the colon : is shorthand for "through" in a sequence) the number of rows (`nrow`) of the `rse` data set. The id variable is added to the end of the `rse` data set (i.e., it becomes the last column).

To remove a variable from a data set, we can assign the value of NULL to the variable:

```
rse$id <- NULL
```

Data frames can be indexed like a matrix. Matrices are indexed by specifying a row number, followed by a comma, and then a column number. For example, to extract the value that corresponds to row 2 and column 4, we would use:

```
rse[2, 4]
```

```
[1] 2
```

Similarly, to print columns 3 through 5 for row 2:

```
rse[2, 3:5]
```

```
  Q3 Q4 Q5
2  0  2  1
```

The code `3:5` is shorthand for 3, 4, and 5 (as we mentioned above). This could also be specified using the combine (c) function:

```
rse[2, c(3, 4, 5)]
```

Columns can also be indexed by specifying the name of the variable. To view the fourth row's age, we would use

```
rse[4, "age"]
```

```
[1] 36
```

or,

```
rse$age[4]
```

```
[1] 36
```

To look at the responses to all the variables for examinees from Taiwan (TW):

```
rse[rse$country == "TW",]
```

	Q1	Q2	Q3	Q4	Q5	Q6	Q7	Q8	Q9	Q10	gender	age	source
424	3	3	3	3	3	3	2	2	3	3	2	55	1
546	3	2	0	2	2	2	2	2	1	1	1	20	1
863	2	2	3	2	2	3	3	2	3	3	1	31	1

	country	person
424	TW	424
546	TW	546
863	TW	863

The == is a binary operator that evaluates to TRUE when the country is TW and FALSE when the country is not TW. The above code selects all the rows that evaluate country == "TW" to TRUE. To continue filtering our data, we examine only the examinees from Taiwan who are under 35 using the & sign (which is a logical operator that stands for "and"):

```
rse[rse$country == "TW" & rse$age < 35,]
```

	Q1	Q2	Q3	Q4	Q5	Q6	Q7	Q8	Q9	Q10	gender	age	source
546	3	2	0	2	2	2	2	2	1	1	1	20	1
863	2	2	3	2	2	3	3	2	3	3	1	31	1

	country	person
546	TW	546
863	TW	863

If we notice an error in our data, we can change it using indexing. For example, if we realized that examinee 424's age (i.e., row 424) should actually have been 35 and not 55, we could update this value as

```
rse[rse$person == 424, "age"]
rse[rse$person == 424, "age"] <- 35
rse[rse$person == 424, "age"]
```

```
[1] 55
[1] 35
```

Data are often stored in a wide (tidy) format (i.e., each row corresponds to a single subject's measurements on multiple items or variables), but for some of the statistical analyses presented in this handbook, we need to convert the data to a long format (i.e., where each row corresponds to a single measurement for a single subject, such that each subject has multiple rows of data corresponding to the number of measurements).

In the following example, we use the **reshape** function to convert the **rse** data set from a wide to a long format. We specify the target direction, the variables that contain the measurements (Q1 through Q10) which are located in columns 1 through 10 in **rse**, the name of the new variable that will contain the multiple measurements for each subject (**timevar = "question"**), the name of the new outcome variable (**v.names = "response"**), and finally the identifier, which is person in the **rse** data set. We save this new data set as **rse_long**. To see the new data format, we print the first six rows of the data using the **head** function.

```
rse_long <- reshape(data = rse,
                    direction = "long",
                    varying = 1:10,
                    timevar = "question",
                    v.names = "response",
                    idvar = "person")
head(rse_long)
```

	gender	age	source	country	person	question	response
1.1	1	10	1	IN	1	1	2
2.1	1	16	1	US	2	1	2
3.1	2	17	1	NL	3	1	1
4.1	1	36	3	GB	4	1	2
5.1	1	15	3	AU	5	1	2
6.1	2	40	3	US	6	1	3

We can convert **rse_long** back to wide format using the same **reshape** function as:

```
rse_wide <- reshape(data = rse_long,
                    direction = "wide",
                    idvar = "person",
                    timevar = "question",
                    v.names = "response")
```

It is sometimes useful to sort data in a long format by an identifier and then by the variable we passed to the **timevar** argument. This is quite helpful especially when the data are in a longitudinal format.

```
rse_long <- rse_long[order(rse_long$person, rse_long$question),]
head(rse_long)
```

	gender	age	source	country	person	question	response
1.1	1	10	1	IN	1	1	2
1.2	1	10	1	IN	1	2	2

1.3	1	10	1	IN	1	3	3
1.4	1	10	1	IN	1	4	3
1.5	1	10	1	IN	1	5	2
1.6	1	10	1	IN	1	6	2

To learn more about data manipulation and basic R syntax, we recommend our readers to review the vignette by John Verzani.[12]

1.5.5 Descriptive and Inferential Statistics

The summary function is a useful way to obtain descriptive statistics about the data. We can apply the summary function to either a data set:

```
summary(rse)
```

or a particular variable (e.g., Q1) in the data set:

```
summary(rse$Q1)
```

Min.	1st Qu.	Median	Mean	3rd Qu.	Max.
0.000	2.000	2.000	2.002	3.000	3.000

In this example, the summary function returns the minimum value, 1st quartile, median, mean, 3rd quartile, and maximum value for Q1.

For variables that are characters or factors, the table function can be more useful to summarize the data as a frequency table:

```
table(rse$country)
```

A2	AE	AL	AR	AU	BA	BE	BG	BH	BR	CA	CH	CL	CR	CY
2	2	1	1	51	1	2	1	1	5	56	2	1	1	1
CZ	DE	DK	DO	EE	EG	ES	ET	FI	FR	GB	GE	GR	HK	HR
3	10	2	1	3	2	6	2	11	6	139	1	1	9	1
HT	ID	IE	IN	IQ	IR	IT	JE	JM	JO	JP	KE	KR	LT	MT
1	3	6	25	2	1	4	1	3	2	2	2	5	4	1
MX	MY	NG	NL	NO	NP	NZ	PE	PH	PK	PL	PS	PT	RO	RS
5	3	1	7	4	1	6	1	19	6	3	1	4	2	3
RU	SA	SE	SG	SI	SK	TH	TR	TT	TW	UA	UG	US	VE	ZA
4	1	9	14	1	1	2	5	2	3	2	1	489	1	11
ZM														
1														

The mean, variance, and standard deviation of a variable can be calculated using the mean, var, and sd commands:

---

[12]https://cran.r-project.org/doc/contrib/Verzani-SimpleR.pdf

```
mean(rse$Q1)
var(rse$Q1)
sd(rse$Q1)
```

```
[1] 2.002
[1] 0.7407367
[1] 0.8606606
```

The `cor` function creates a correlation matrix. To use this function, all variables must be numeric. Below, we use the `subset` function to select variables Q1 through Q10 in the `rse` data set,[13] create a correlation matrix for the new data set, which we save as `rse_cor`, and then round the correlations to the nearest hundredth (the 2 stands for the number of decimal places we want) when we print the correlation table:

```
rse_sub <- subset(rse, select = Q1:Q10)
rse_cor <- cor(rse_sub)
round(rse_cor, 2)
```

	Q1	Q2	Q3	Q4	Q5	Q6	Q7	Q8	Q9	Q10
Q1	1.00	0.69	0.49	0.58	0.47	0.61	0.56	0.35	0.39	0.49
Q2	0.69	1.00	0.45	0.53	0.50	0.57	0.52	0.29	0.39	0.46
Q3	0.49	0.45	1.00	0.45	0.63	0.59	0.59	0.41	0.56	0.61
Q4	0.58	0.53	0.45	1.00	0.39	0.48	0.48	0.28	0.37	0.41
Q5	0.47	0.50	0.63	0.39	1.00	0.55	0.56	0.38	0.52	0.57
Q6	0.61	0.57	0.59	0.48	0.55	1.00	0.74	0.47	0.52	0.61
Q7	0.56	0.52	0.59	0.48	0.56	0.74	1.00	0.47	0.52	0.58
Q8	0.35	0.29	0.41	0.28	0.38	0.47	0.47	1.00	0.51	0.53
Q9	0.39	0.39	0.56	0.37	0.52	0.52	0.52	0.51	1.00	0.74
Q10	0.49	0.46	0.61	0.41	0.57	0.61	0.58	0.53	0.74	1.00

A covariance matrix can also be requested in a similar way, but by substituting the `cov` function for the `cor` function.

To test the significance of a correlation, we can use the `cor.test` function. For example, if we want to test the significance of the relationship between Q1 and Q2, we can use the following command:

```
cor.test(rse$Q1, rse$Q2)
```

```
Pearson's product-moment correlation

data:  rse$Q1 and rse$Q2
t = 30.486, df = 998, p-value < 2.2e-16
alternative hypothesis: true correlation is not equal to 0
```

---

[13]We could also select only some variables, such as `select=c(Q1, Q6, Q7, Q9)`.

```
95 percent confidence interval:
 0.6608709 0.7251887
sample estimates:
      cor
0.6944142
```

A t-test can be performed using the t.test function. Below, we subset the male and female examinees from the **rse** data set and see if they differ on age. In the **subset** function, we use the | symbol, which is a logical operator that corresponds to "or." This helps us select examinees whose gender was either 1 or 2 in the **rse** data set. We save the new data set as **rse_gender** and use it with the **t.test** function. From the output below, we find that male and female examinees do not differ on age ($p = 0.6092$).

```
rse_gender <- subset(rse, gender == 1 | gender == 2)
t.test(age ~ gender, data = rse_gender)
```

```
Welch Two Sample t-test

data: age by gender
t = -0.51135, df = 808.51, p-value = 0.6092

alternative hypothesis:
true difference in means is not equal to 0

95 percent confidence interval:
 -10.929294    6.411843

sample estimates:
mean in group 1 mean in group 2
       29.57825        31.83697
```

Simple and multiple linear regression analyses can be performed with the lm function. Below, we regress Q1 onto Q2 and Q3, save the results to **mod1**, show what is contained within **mod1**, and print a summary of the results.

```
mod1 <- lm(Q1 ~ Q2 + Q3, data = rse)
names(mod1)
summary(mod1)
```

```
 [1] "coefficients"  "residuals"      "effects"
 [4] "rank"          "fitted.values" "assign"
 [7] "qr"            "df.residual"   "xlevels"
[10] "call"          "terms"         "model"

Call:
lm(formula = Q1 ~ Q2 + Q3, data = rse)
```

```
Residuals:
    Min       1Q    Median       3Q      Max
-2.88984 -0.24594  0.04116  0.32456  1.82676

Coefficients:
            Estimate Std. Error t value Pr(>|t|)
(Intercept)  0.31494    0.05419   5.812 8.31e-09 ***
Q2           0.64390    0.02654  24.260  < 2e-16 ***
Q3           0.21440    0.02256   9.504  < 2e-16 ***
---
Signif. codes:
0 '***' 0.001 '**' 0.01 '*' 0.05 '.' 0.1 ' ' 1

Residual standard error: 0.5936 on 997 degrees of freedom
Multiple R-squared:  0.5252,Adjusted R-squared:  0.5243
F-statistic: 551.5 on 2 and 997 DF,  p-value: < 2.2e-16
```

When we run a regression analysis, as with many functions in R, the output object contains a variety of information, which we can see using the **names** or **str** functions. For regression, this includes the coefficients, residuals, predicted values, and more. Above, we find that Q2 and Q3 are significant predictors of Q1, that R^2 is 0.525, and that the residual standard error, $\hat{\sigma}$, is 0.594. Predicted values and residuals from this model can be obtained using the following commands:

```
pred_values <- predict(mod1)
resid_values <- resid(mod1)
```

We can request R's default diagnostic plots for evaluating regression models using the **plot** function. Prior to doing this below, we tell R to partition the graphic window into a 2 x 2 grid using the **par** command with the **mrow** argument. The other argument, **mar**, sets the size of the margins around the plots. These two arguments allow us to plot all four diagnostic plots simultaneously and trim away some of the white space between them. In Figure 1.2, the top-left plot is a residual plot, useful for detecting patterns in residuals and non-constant variance; the top-right plot is a normal Q–Q plot, useful for assessing normality assumption; the bottom-left plot is a scale-location plot, useful for detecting non-constant variance; and the bottom-right plot is a leverage plot, useful for identifying influential cases.

```
par(mfrow = c(2, 2), mar=c(2, 4.1, 2, 2))
plot(mod1)
```

We see evidence of non-constant variance in Figure 1.2 as well as violations of normality (note the heavy tail). If this was a model that we were interested

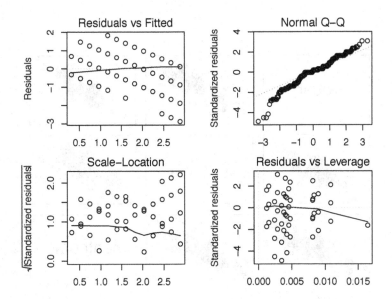

FIGURE 1.2
Default diagnostic plots from a multiple regression model.

in, we would have to consider either transformations for the variables or a different method of estimation (e.g., weighted least squares).

1.5.6 Plotting in R

In this handbook, we use the basic R graphics functions and the functions in **lattice** (Sarkar, 2008). In the following sections, we briefly show how to create basic plots in R and a trellis plot in **lattice**.

1.5.6.1 Base R Graphics

A scatter plot can be created using the `plot` function. We demonstrate this by plotting the vocab against the mathmtcs variable in the **interest** data set:

```
plot(vocab ~ mathmtcs, data = interest)
```

We can also add colors to the plot based on the examinees' gender:

```
plot(vocab ~ mathmtcs, data = interest, col = gender)
```

and add a LOWESS smoother (Cleveland, 1985) to this plot.

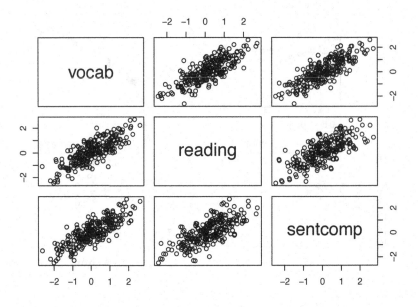

FIGURE 1.3

Scatter plot matrix of verbal measures in the interest data set.

```
plot(vocab ~ mathmtcs, data = interest, col = gender)
lines(lowess(interest$mathmtcs, interest$vocab))
```

A scatter plot matrix can be created using the `pairs` function. Because there are 33 variables in the `interest` data set, this would create a very large scatter plot matrix. In the `interest` data, the vocab, reading, and sentcomp variables correspond to a verbal measure of intelligence quotient. Below we subset these variables and create a scatter plot matrix of this new data set, which is shown in Figure 1.3.

```
verbal <- subset(interest, select = c(vocab, reading, sentcomp))
pairs(verbal)
```

Histograms can be created using the `hist` function:

```
hist(interest$vocab)
```

Box plots can be created with the `boxplot` function:

```
boxplot(interest$vocab)
```

To create a box plot by a grouping variable, for example, by gender, we would do the following:

```
boxplot(interest$vocab ~ interest$gender)
```

Finally, a stem and leaf plot can be created using stem:

```
stem(interest$vocab)
```

```
 The decimal point is at the |

 -2 | 6
 -2 | 200
 -1 | 99888877666655
 -1 | 4333322222211111000
 -0 | 999999888888887777776666666655555555
 -0 | 4444444433333333222222222222222211111111110000000
  0 | 0000011111111222222222233333333344444444
  0 | 555555556666777777788888999999999
  1 | 000000001111111111222222333333444
  1 | 555566667778999
  2 | 0111123
  2 | 6
```

1.5.6.2 Lattice Graphics

We will principally use the **lattice** package to make trellis plots. To create a scatter plot with **lattice**, we need to use the xyplot function:

```
xyplot(vocab ~ reading, data = interest)
```

To create a trellis plot, we use the | symbol, which allows us to create a plot of vocab by reading conditional on a third variable. Below, we create a trellis plot by gender.

```
xyplot(vocab ~ reading | gender, data = interest)
```

In Figure 1.4 we see that the relationship between vocab and reading does not seem to depend on gender. In other words, there is no interaction between reading and gender. To color the points by gender, the group = gender argument must be included in the xyplot function.

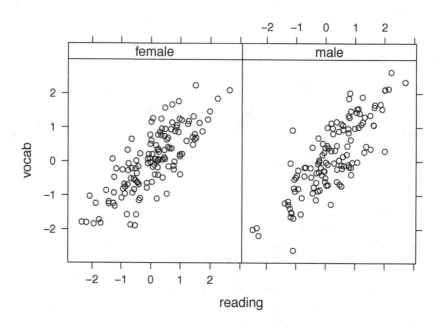

FIGURE 1.4
Scatter plot of vocab by reading conditional on gender.

1.6   Installing Packages Used in This Handbook

Before we conclude this chapter, it is important that all of the packages used in this handbook are installed. Below, we show the commands that our readers should run to make sure all the packages for this handbook are installed.

```
install.packages("boot")
install.packages("difR")
install.packages("equate")
install.packages("faoutlier")
install.packages("GPArotation")
install.packages("lattice")
install.packages("lavaan")
install.packages("lme4")
install.packages("mirt")
install.packages("psych")
install.packages("semPlot")
install.packages("shiny")
```

```
install.packages("devtools")
devtools::install_github("cddesja/hemp")
```

If our readers wish to cite R and the R packages in their publications, they can use the `citation` function in R to view the proper citations. To view the citation for R, the following command should be used:

```
citation()
```

The `citation` function can also be used for viewing the citation for a particular R package:

```
citation("package name")
```

where "package name" should be replaced with the name of the package that needs to be cited (e.g., `citation("lme4")` for the **lme4** package).

Finally, the readers who prefer BibTeX for managing their references can use the following commands to view the citations for R and the R packages:

```
toBibtex(citation())
toBibtex(citation("package name"))
```

where "package name" should be replaced with the name of the package of interest.

1.7 Chapter Summary

In this chapter, we showed how to install R and how to perform basic tasks in R, such as reading in data, cleaning data, descriptive and inferential statistics, and plotting. We also showed how to install the necessary packages to be able to follow the examples presented throughout this handbook. For readers looking for a more in-depth introduction to R, we recommend *Introduction to R* by Venables and Smith (2016) and the introduction by Verzani (Verzani, 2002), and for the readers looking for an exhaustive reference on R, we strongly recommend *The R Book* by Crawley (2013).

2

Classical Test Theory

2.1 Chapter Overview

In this chapter, we introduce educational measurement and psychometrics using the classical test theory (CTT) framework. We begin by defining measurement, tests, and scales of measurement, and then move into reliability and validity in the context of CTT. After a brief introduction, we present several examples of how to calculate many of the CTT-based statistics in R. In addition, we introduce bootstrapping as an option to obtain confidence intervals for CTT-based parameters when their sampling distributions are unclear or assumptions are violated. We finish the chapter by presenting item analysis. The R functions used in this chapter come from the **hemp** and **boot** packages.

2.2 What Is Measurement?

Generally speaking, measurement is the quantification of a construct by assigning numerical values or qualitative labels based on a set of rules, principles, or operations. For example, Mr. Williams, an eighth grade math teacher, wants to measure his students' knowledge of algebra. To accomplish this task, Mr. Williams decides to design and use a math test that measures his students' knowledge of algebra. During the development of his math test, Mr. Williams encounters several measurement challenges:

- What concepts in algebra should my students know?

- How many questions should I use?

- How many questions should I include per concept area, and what should be the format of the questions?

- Do I need to write new questions or can I use questions from a pre-existing test?

- Do I need to create multiple forms because of the possibility of cheating? If so, how can I ensure they are of the same difficulty or comparable?

- How can I be sure that my questions are really measuring algebra knowledge and not something else?

- How should the test be scored?

- Should each question be scored in the same manner? For example, should each question receive the same number of points? Should students receive points only if the question is fully correct or can they receive points if it is partially correct?

Measurement instruments such as tests, scales, surveys, and questionnaires are the typical vehicles that we, as teachers, educators, psychologists, practitioners, or researchers, use to measure a construct, trait, or domain of interest. We construct measurement instruments with a specific purpose in mind, such as developing a math test to measure algebra knowledge. A construct, trait, or domain is a theoretical entity or concept, such as reading ability, intelligence, or executive functioning. In education and psychology, we usually cannot measure constructs directly because they are latent (unobserved), and thus we must make inferences about them from their manifestations. To measure a latent construct, we develop and use items (or tasks) in measurement instruments. These items are typically referred to as "manifest variables" that provide an operational definition of the latent construct being measured. Latent constructs are sometimes broadly defined, such as intelligence, or may be more narrowly defined, such as knowledge of trigonometric functions in math. The distinction between terms - such as a test, an instrument, and a survey - is unimportant in this handbook, as the measurement frameworks and concepts that we will explain can be generalized to all types of measurement tools. Therefore, these terms will be used interchangeably throughout this handbook for the sake of simplicity.

2.3   Issues in Measurement

Important topics in educational measurement and psychometrics include validity, reliability, invariance (i.e., fairness), and scaling (De Ayala, 2013). To date, measurement issues related to these topics have been the main focus of the literature on educational measurement and psychometrics. Validity refers to the degree to which we are measuring what we have intended to measure. Validity aims to investigate whether the manifest variables are the true manifestations of the target construct or something else. Validity involves collecting evidence and building a case supporting the use of a measurement instrument.

All validity evidence essentially falls under the umbrella of construct validity.[1] Chapter 4 covers factor analysis, which is often considered a method for assessing construct validity.

Although it has always been considered a stand-alone concept, reliability is, in fact, a form of validity evidence. Reliability refers to the consistency of the results from a measurement instrument. For example, if we administered an instrument repeatedly, would the results be the same, similar, or would they vary significantly? If the scores vary significantly, then they will have low reliability and contain a high amount of measurement error. To make valid inferences based on the results from a measurement instrument, the reliability of the results is essential. In other words, a measurement instrument cannot yield valid inferences unless the obtained results are highly reliable. However, high reliability alone is not sufficient for making valid inferences. For example, a knowledge test to obtain a driving license can be a highly reliable instrument. However, if the developers of this test used the results to decide who could be a race car driver, this measurement process would yield invalid inferences.

Generally speaking, we want the results or scores from an instrument to reflect what we intend to measure. If the results are independent of examinee characteristics (e.g., gender, ethnicity, and cultural background) and test administrations (e.g., the administration of a test in 2016 and 2017), then we can say that the instrument is invariant. Violations of invariance may jeopardize the validity of the inferences made based on the results. Many of the psychometric models considered in this book assume invariance or involve explicitly testing this assumption. One form of invariance, measurement invariance, is covered in detail in Chapter 11.

Finally, a score scale underlying a measurement instrument has important properties. A scale typically consists of a set of values or labels that are used to categorize or quantify what is being measured. In the algebra test example, a continuous score scale can be used for assigning a total test score between 0 and 100 or a numerical value between 0 and 4; an ordinal scale can be used for assigning a letter grade, such as an A through an F; and a nominal scale can be used for assigning a categorical label, such as pass or fail, proficient or not proficient. In the following section, we briefly explain how to decide what type of measurement scale should be used.

2.3.1 Type of Scales

The nature of a scale is very important in measurement as this has implications for the selected methods and models. Because the models presented

---

[1] Some authors, such as Thorndike and Thorndike-Christ (2010), also present validity separately as content, criterion-related, and construct validity, but this is unimportant for our purposes.

throughout this handbook depend on the scale, a more thorough discussion of measurement scales is necessary. Measurement scales are typically classified as nominal, ordinal, interval, and ratio. A nominal scale consists of qualitative and unordered categories. For example, a scale that assigns numbers to someone's favorite color is inherently unordered, and by extension qualitative, because we cannot rank color.[2] For example, a "1" may be assigned to red, a "2" to blue, a "3" to green, and so on. However, it would be improper to interpret this scale as if the assigned values are quantitative - such as green is greater than blue and red or that green is three times greater than red. Similarly, a scale that assigns numbers based on someone's preferred gender pronoun or ethnicity is inherently unordered, and thus nominal.

Unlike nominal scales, ordinal scales are inherently ordered. An ordinal scale can be either quantitative (e.g., first, second, and third positions in a race) or qualitative (e.g., excellent, proficient, developing, and beginning in a scoring rubric). However, with ordinal scales, we do not know the distance between adjacent categories and there is no reason to assume that the distance between adjacent categories is equivalent. A typical example of an ordinal scale is a Likert-type item that asks about the level of agreement to some topic. Consider the following statement: I enjoy solving algebra problems. This item could be scored as "1" for strongly disagree, "2" for disagree, "3" for neither agree nor disagree, "4" for agree, and "5" for strongly agree. It is possible to imagine that people range in their underlying enjoyment of algebra and that their enjoyment is in fact continuous. However, by virtue of our instrument, their enjoyment is discretized and people are forced to respond in a manner consistent with our scale. There is no reason to believe that the distance in "enjoyment in solving algebra problems" is the same between someone responding with a disagree and someone responding with neither an agree nor a disagree, or someone responding with a disagree and someone responding with a strongly disagree. Although the numerical values assigned to ordinal levels may imply that the distance between the levels is equal to 1 point, these values cannot be compared quantitatively as they have been chosen arbitrarily. For example, we might have assigned a "1" to strongly disagree, a "3" to disagree, a "4" to neutral, a "5" to agree, and a "7" to strongly agree. This would change the distance between the adjacent categories; however, the ordering would remain the same. Therefore, when we are dealing with ordinal scales, we must exploit the ordered nature of the scale but realize that the assigned numbers are nothing more than arbitrary labels.

If we have an ordered scale with equal intervals, then our scale is said to be at least an interval scale. If the scale has an absolute zero, then our scale would be a ratio scale, otherwise it is an interval scale. The classic example of an interval scale is temperature recorded on the Celsius or Fahrenheit scale.

---

[2]Technically, we could rank colors by wavelength, but this ordering makes little sense in this example.

0 degrees on the Celsius scale would not imply the absence of temperature or heat. However, the distance between two adjacent degrees is always the same on the Celsius scale. Ratio scales also have the equal distance feature as interval scales. In addition, ratio scales have a real zero point, indicating the absence of the construct being measured. Unlike the ordinal and interval scales, ratio scales allow interpreting the proportions of the values on the scale (e.g., 10 inches are twice as long as 5 inches).

Data from nominal and ordinal scales are often referred to as categorical or discrete data, while data from interval and ratio scales are referred to as continuous data. With continuous data, we sometimes make an assumption that the data are normally distributed either marginally or conditionally based on a set of predictors, covariates, or independent variables. With interval and ratio scales, we can calculate descriptive statistics (e.g., mean and standard deviation); however, such statistics are inappropriate with nominal and ordinal scales. We may examine scatter plots, histograms, or stem and leaf plots with interval and ratio scales. With nominal or ordinal scales, we may construct contingency tables to show the number or proportion of respondents endorsing a certain response, and create bar charts, dot plots, or mosaic plots.

To briefly show the differences in how to summarize categorical and continuous data, we use the `interest` data set in the **hemp** package. The `interest` data come from a fabricated cognitive, personality, and vocational interest inventory. For more information about this data set, `?interest` can be run in the R console. To enable the functions and data sets in the **hemp** package, we need to use the `library` command as follows:

```
library("hemp")
```

Once the **hemp** package has been activated, the `interest` data set becomes available to use. One of the variables in the `interest` data set is gender. The gender variable was originally coded as 1 for female respondents and 2 for male respondents. We begin by recoding this numerical variable into a categorical variable with the actual gender labels using the `ifelse` function. The `ifelse` function defines a conditional operation, such as "if this statement is true, then do ..., otherwise do ...." In the example below, we use `ifelse` to find and recode the observations (i.e., respondents) for which the gender variable is equal to 1. When the condition `gender==1` is satisfied, we assign the label "female"; otherwise, we assign the label "male." Finally, we save this new variable as `gender_nominal`.

```
gender_nominal <- ifelse(interest$gender == 1, "female", "male")
```

Next, we transform age, which is a continuous, numerical variable in the `interest` data set, into an ordinal variable for didactic purposes. Although transforming age into an ordinal variable is likely to result in information loss, we want to demonstrate how a continuous variable on a ratio scale can be converted to an ordinal variable. To transform age into an ordinal variable, we use the `cut` function to create bins that are a width of 10 years for the ages 10 to 70 years. The first bin contains examinees that are 10 to 19 years old, the second bin contains examinees that are 20 to 29 year olds, and so on. For R to know that this new variable is an ordinal variable and not a nominal variable (which R assumes by default), we use the `ordered` function and save it as `age_ordinal`.

```
age_nominal <- cut(interest$age,
          breaks = seq(10, 70, by = 10))
age_ordinal <- ordered(age_nominal)
```

To summarize our new variables, we use the `table` function.

```
table(gender_nominal)
```

```
gender_nominal
female    male
   128     122
```

```
table(age_ordinal)
```

```
age_ordinal
(10,20] (20,30] (30,40] (40,50] (50,60] (60,70]
      9      34      89      88      23       7
```

From the printed output above, we see that there are slightly more females than males and that a majority of examinees are between 30 and 50 years of age. Using the `ftable` function, we can also create contingency tables with multiple nominal and ordinal variables.

```
ftable(age_ordinal, gender_nominal)
```

```
            gender_nominal female male
age_ordinal
(10,20]                         4    5
(20,30]                        19   15
(30,40]                        51   38
(40,50]                        39   49
(50,60]                        11   12
(60,70]                         4    3
```

The contingency table returned from the `ftable` function shows that both males and females are mostly between 30 and 50 years of age. In addition to a frequency table using the `table` function, we can also create a proportion table using the `prop.table` function. The `prop.table` creates a proportional table based on a frequency table derived from the `table` function. To create a proportion table for our ordinal age variable, we first create a frequency table and pass this table as an argument in the `prop.table` function.

```
age_table <- table(age_ordinal)
prop.table(age_table)
```

```
age_ordinal
(10,20] (20,30] (30,40] (40,50] (50,60] (60,70]
 0.036   0.136   0.356   0.352   0.092   0.028
```

The output from the `prop.table` shows that approximately 70% of the examinees in the **interest** data set are between 30 years old and 50 years old. Like the **table** function, the `prop.table` function can also be applied to contingency tables.

```
prop.table(table(age_ordinal, gender_nominal))
```

```
            gender_nominal
age_ordinal female  male
    (10,20]  0.016 0.020
    (20,30]  0.076 0.060
    (30,40]  0.204 0.152
    (40,50]  0.156 0.196
    (50,60]  0.044 0.048
    (60,70]  0.016 0.012
```

In the examples presented so far, we again see an important feature of R that everything can be saved and passed on to future functions. This allows us to create code that is easier to read without needing to nest functions in functions[3].

To examine nominal and ordinal variables visually, we can create barcharts and dot plots using the **barchart** and **dotplot** functions in the **lattice** package (Sarkar, 2008). For example, to create a barchart, we can run the following code:

```
library("lattice")
barchart(age_ordinal)
```

We imagine many of our readers may be unfamiliar with dot plots. We find dot plots extremely insightful for uncovering patterns in categorical data (see Cleveland (1985) and Cleveland (1993) for illustrative examples). They are a little more difficult to create because they require the frequency of the categories in a nominal variable. Fortunately, this information is already contained in the **age_table** object and we can just pass this object to the **dotplot** function as follows:

---

[3]An alternative is to use the pipe operator, %>%, which was originally created for the **magrittr** package (Bache & Wickham, 2014). The pipe operator allows users to string together functions without needing to write nested functions and obfuscatory code in R.

```
dotplot(age_table)
```

Another extremely powerful plot is a type of plot known as a trellis or facet plot. A trellis plot can show the relationship between two variables conditional on a third (as we briefly mentioned in Chapter 1). For example, if we want to see how the age distribution for the examinees differed by gender, we would do the following:

```
age_gender_table <- table(gender_nominal, age_ordinal)
age_gender_df <- data.frame(age_gender_table)
dotplot(age_ordinal ~ Freq | gender_nominal,
          age_gender_df, xlab = "Frequency", ylab = "Age")
```

The above code first creates a two-way table (`age_gender_table`) and then converts it into a data frame (`age_gender_df`). This is a necessary, intermediate step for plotting with the `dotplot` function. As a reminder the "|" sign can be interpreted as "conditional on" and not as a boolean "or." In the example above, the `dotplot` function plots the frequency distribution of the age variable conditional on the gender variable.

Figure 2.1 shows the result of executing this code.[4] We see in the dot plot that the age distributions are similar between females and males except that the mode for males is between 40 and 50, while for females it is between 30 and 40, which implies that the female examinees are slightly younger in our sample than the male examinees.

Next, we consider the vocab variable that corresponds to a vocabulary test in the `interest` data set. This variable is on an interval scale. Summary data for this variable can easily be obtained with the `summary` function.

```
summary(interest$vocab)
```

Min.	1st Qu.	Median	Mean	3rd Qu.	Max.
-2.62000	-0.60500	0.04000	0.09016	0.86000	2.63000

We see that the minimum score on the vocabulary test was -2.62 and the maximum score was 2.63. The mean is .09. The 1st and 3rd quartiles, as well as the median, are also provided by the `summary` function.

With interval or ratio scales, it is often of interest to examine the extent to which a variable is normally distributed. While there are a variety of statistics and plots available for assessing the normality assumption, a particularly useful option is the Q–Q plot (Faraway, 2014). We prefer the Q–Q plot to histograms for assessing normality because of the arbitrariness of bin width and numbers. We strongly recommend against calculating skewness and kur-

---

[4]Note that the colors in your dot plot will be different than ours as we used the grayscale for this handbook.

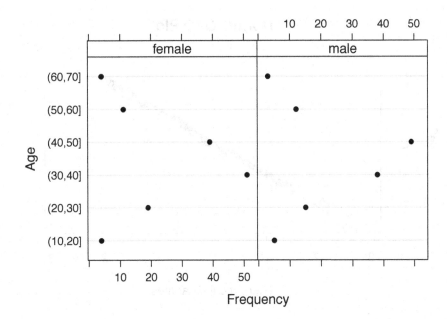

FIGURE 2.1
Dot plot of age by sex of examinees taking the interest inventory.

tosis statistics because it is generally more useful to see the distributions of the variables and to understand the nature of the deviation.

The code to construct the Q–Q plot is shown below and the resulting plot is shown in Figure 2.2. We use the `qqnorm` function to create the Q–Q plot and then the `qqline` to overlay a line that shows where the points would fall if the distribution were perfectly normal. In the plot, we see that there are some slight deviations at the tails but no major violations of the normality.

```
qqnorm(interest$vocab, ylab = "vocab")
qqline(interest$vocab)
```

While it is beyond the scope of this handbook, we want to note that the type of scale affects the type of statistical tests that should be considered. Tests appropriate for interval or ratio scale data (e.g., t-tests, linear regressions) are generally not appropriate for categorical data. We strongly refer readers unfamiliar with categorical data analysis to Agresti (2002) and Faraway (2016). Many of the methods presented throughout this book are appropriate for all scales, while some are specific to a certain scale. Care will be taken throughout to make it clear when a method requires a certain measurement scale.

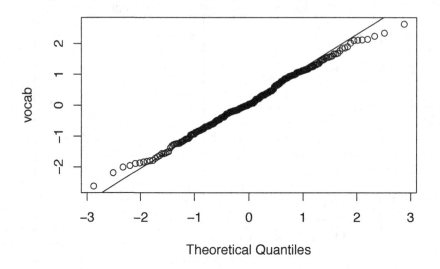

FIGURE 2.2
Q–Q plot of the vocab variable in the interest data set.

2.4   The Classical Test Theory Framework

Classical test theory (CTT), also called true score theory, is a measurement framework that allows us to understand, manipulate, and interpret measured outcomes. The premise of CTT is that every measurement contains error and that any, and all, observations are imperfect. The CTT model decomposes an observed score from a measurement instrument into true score and error components. Mathematically, CTT is expressed as:

$$X = T + E. \tag{2.1}$$

In the CTT model, X is the observable measurement/test score; T is the true (latent) measurement/total test score; and E is the random error. When focus is on the individual items rather than the unobservable total score, then the common factor model (presented in Chapter 4) or item response theory (introduced in Chapter 5) is more appropriate than the CTT model. To use the CTT model, there are four additional assumptions beyond the general form presented in Equation 2.1:

1. $E(X) = T$, the expected value of the observed score is the true score,

2. $Cov(T, E) = 0$, the true score and error are independent,

3. $Cov(E_1, E_2) = 0$, the errors across test forms are independent, and

4. $Cov(E_1, T_2) = 0$, the error on one form is independent of the true score on another form.

Because of these assumptions, the CTT model can be re-expressed as the simple sum of orthogonal (i.e., uncorrelated) variance components as shown below:

$$\sigma_X^2 = \sigma_T^2 + \sigma_E^2. \tag{2.2}$$

Equation 2.2 states that observed score variance is the sum of true score and error variances. Within this model, true score variance is assumed to be constant (e.g., it will never change regardless of the form of the instrument, the date of the assessment, etc.), while error variance fluctuates (e.g., some forms may contain more error than others). The measurement error can be divided into random (unpredictable and inconsistent) and systematic (constant and predictable) error.

2.4.1 Reliability

Reliability is the proportion of observed score variance that can be attributed to true score variance. From Equation 2.2, an index of test reliability can be derived as follows:

$$\text{reliability} = \frac{\sigma_T^2}{\sigma_X^2}. \tag{2.3}$$

There are four general types of reliability: test-retest (coefficient of stability), parallel forms (coefficient of equivalence), alternate forms (alternate forms reliability), and internal consistency. Here focus is on internal consistency as examinees taking multiple forms or taking the same form on multiple administrations are typically rare in practice. In many cases, the other forms of reliabilities can be estimated using either a Pearson or tetrachoric correlation with some correction for attenuation if there is range restriction.

In the following example, we use the SAPA data set to demonstrate how to calculate various estimates of internal consistency in R. The SAPA data set, originally included in the **psych** package (Revelle, 2017), has been repackaged and scored for our readers' convenience in the **hemp** package. The instrument consists of 1525 responses to a 16-item multiple-choice ability test taken from a web-based personality assessment project called the Synthetic Aperture Personality Assessment (SAPA).[5] The first four items measure basic reasoning, the second four items measure alphanumeric series, the third four items measure matrix reasoning, and the final four items measure spatial rotation.

We begin the analysis by calculating the number of missing observations for each item using the num_miss function in **hemp**. The output shows that only letter.58 contains no missing data, while the other items contain one or two missing observations.

```
num_miss(SAPA)
```

	num_miss	perc_miss
reason.4	2	0.13
reason.16	1	0.07
reason.17	2	0.13
reason.19	2	0.13
letter.7	1	0.07
letter.33	2	0.13
letter.34	2	0.13
letter.58	0	0.00
matrix.45	2	0.13
matrix.46	1	0.07
matrix.47	2	0.13
matrix.55	1	0.07
rotate.3	2	0.13
rotate.4	2	0.13
rotate.6	2	0.13
rotate.8	1	0.07

A simple measure of internal consistency is split-half reliability. A split-half reliability estimate can be obtained by splitting a test into two equivalent halves, calculating the total scores on the two halves, and correlating them. There are a myriad of ways to create the split-half (e.g., selecting every other item, randomly, etc.). Below we calculate the split-half reliability by selecting every other item with the split_half function and by specifying type = "alternate".

```
split_half(SAPA, type = "alternate")
```

```
[1] 0.758
```

---

[5]https://sapa-project.org/. For more details, see the **psych** package (Revelle, 2017)

For the `SAPA` data set, the split-half reliability was estimated to be 0.758. We can also calculate the split-half reliability by splitting the test at random by specifying `type = "random"`.

```
set.seed(1)
split_half(SAPA, type = "random")
```

```
[1] 0.717
```

The split-half reliability with a random split resulted in a reliability estimate of 0.717. By specifying `set.seed(1)`, we set the random number generator so that our readers' results will be the same as the result presented above.[6] This is useful for reproducing the same results later. To obtain a new estimate based on a different random split, we can either remove the `set.seed(1)` command or change the seed to a different value (e.g., `set.seed(555)`).

Split-half reliability estimate is known to be downwardly biased (R. J. Cohen, Swerdlik, & Sturman, 2013). The Spearman-Brown correction can be applied to adjust for this. To do this, just pass the `sb = TRUE` argument to the `split_half` function.

```
split_half(SAPA, type = "alternate", sb = TRUE)
```

```
[1] 0.8623436
```

After applying the Spearman-Brown correction, the split-half reliability is now estimated at 0.862, which is quite a bit higher. Given our present reliability estimate, we can also determine the length a test should be to get a desired reliability. We can do this using the `test_length` function in **hemp**. Assuming we desire a reliability of 0.95, we can determine how long our test should be. In the call to `test_length`, we specify `r_type = "split"` in order for the present reliability to be calculated using the split-half reliability with the Spearman-Brown correction. We could also specify a number to `r_type` if we calculated reliability elsewhere (as we did above).

```
test_length(SAPA, r = .95, r_type = "split")
```

```
[1] 49
```

```
test_length(SAPA, r = .95, r_type = .862)
```

```
[1] 49
```

If we want a test with a desired reliability of 0.95, given that our present test has a reliability of 0.862 based on 16 items, we would need a test that consists of at least 49 items.

---

[6] Note that random number generation may work differently across different versions of R, and thus the results may still be different.

The most common measure of internal consistency is coefficient alpha (Cronbach, 1951). Coefficient alpha represents the mean of all possible split-half correlations. It can be calculated using the `coef_alpha` function in **hemp**.

```
coef_alpha(SAPA)
```

```
[1] 0.841
```

The output returned from the `coef_alpha` function shows that the estimate of coefficient alpha is 0.841 for the `SAPA` data set. While point estimates of reliability can be useful, it is generally helpful to have a sense of variation around estimates. Confidence intervals provide a means to quantify a range of plausible values for unknown parameters given a level of confidence (typically 95%). When the sample size is small, modeling assumptions are not met, or a sampling distribution of a parameter is unknown, then bootstrapping can be used to construct an empirical sampling distribution, which we can then use to create confidence intervals (Efron & Tibshirani, 1986).[7]

Our motivation for introducing the bootstrapping is that it allows the creation of confidence intervals and uncertainty regardless of the parameter being estimated. Therefore, while we illustrate the use of the bootstrap for coefficient alpha, this can easily be applied for the split-half reliability above, the validity or item analysis statistics presented later in this chapter, and in many other contexts presented throughout the book.

To perform the bootstrap, we use the **boot** package (Canty & Ripley, 2017). The **boot** package offers a myriad of options for both the parametric and nonparametric bootstrap via the `boot` function. The technical details about these methods are described in A. C. Davison and Hinkley (1997) and are beyond the scope of this handbook. To begin the analysis, we first activate the **boot** package

```
library("boot")
```

Next, we use the `boot` function to implement the bootstrapping procedure. At a minimum, the `boot` function requires a data set, a function to bootstrap, and the number of samples to draw. First, we need to create a function to pass to the `boot` function, which we call `alpha_fun`. This function takes two arguments, a data set, called `data`, and an index matrix, called `row`. These arguments are then passed to the `coef_alpha` function. This will allow the boot function to sample the examinees with replacement from `SAPA` and create an empirical distribution for coefficient alpha (see Canty (2002) for more details).

```
alpha_fun <- function(data, row){
  coef_alpha(data[row, ])}
```

---

[7]Traditional confidence intervals for coefficient alpha are also available from the `alpha` function in the **psych** package.

Below we use bootstrapping with 10,000 draws to create an empirical distribution, save the results as `alpha_boot`, and print the results.

```
alpha_boot <- boot(SAPA, alpha_fun, R = 1e4)
alpha_boot
```

```
ORDINARY NONPARAMETRIC BOOTSTRAP


Call:
boot(data = SAPA, statistic = alpha_fun, R = 10000)


Bootstrap Statistics :
    original      bias     std. error
t1*    0.841 -0.0003296 0.005473752
```

In the output, `t1*` represents coefficient alpha and `original` is the original estimate using all the data, `bias` is the mean of the empirical distribution minus this original estimate, and `std. error` is the standard error of our estimator.

Figure 2.3 shows a histogram and Q–Q plot of the empirical distribution for coefficient alpha based on the 10,000 samples. The distribution is roughly normal, with the exception of slight deviations at the tails. This implies that using the bootstrap normal intervals, which assume an asymptotic normal distribution to construct confidence intervals, may be appropriate.

```
plot(alpha_boot)
```

Further, we use the `boot.ci` function to calculate 95% confidence intervals using bootstrap normal, basic bootstrap, bootstrap percentile, and adjusted bootstrap percentile (BCa) intervals. Briefly, basic, percentile, and BCa intervals make fewer assumptions than the bootstrap normal intervals (i.e., no asymptotic normality assumption), with the BCa interval behaving better asymptotically than the basic and percentile intervals. If the empirical distribution departs from normality, then the basic, percentile, and BCa intervals are a better choice. A more complete description of the difference between these intervals can be found in detail in A. C. Davison and Hinkley (1997), and the important point for our example is that the intervals agree to the hundredths, with the bootstrap normal intervals having slightly greater precision.

```
boot.ci(alpha_boot, type = c("norm", "basic", "perc", "bca"))
```

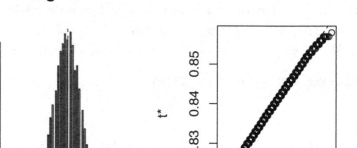

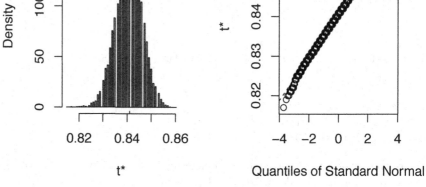

FIGURE 2.3
Empirical distribution for coefficient alpha (n = 10,000).

```
BOOTSTRAP CONFIDENCE INTERVAL CALCULATIONS
Based on 10000 bootstrap replicates

CALL :
boot.ci(boot.out = alpha_boot, type = c("norm", "basic", "perc",
    "bca"))

Intervals :
Level       Normal                  Basic
95%   ( 0.8306,  0.8521 )    ( 0.8310,  0.8520 )

Level      Percentile               BCa
95%   ( 0.830,  0.851 )    ( 0.830,  0.851 )
Calculations and Intervals on Original Scale
```

We conclude that for most purposes our reliability appears to be high enough not to necessitate the additional burden to the examinees of increasing the length of the instrument, and we would be comfortable making most decisions with this estimate of reliability (provided the decisions are not extremely high stakes).

2.4.2 Validity

Validity evidence takes many forms, and quantifying validity evidence is fairly straightforward with R. A common form of validity evidence is expert opinion. The opinions of experts can help comment on the appropriateness of the content of the items, whether the instrument is adequately sampling all the concepts within the construct, and whether the items are essential for measuring the target construct. One way to quantify the latter is with the content validity ratio, CVR (Lawshe, 1975). The CVR is defined as:

$$CVR = \frac{n_e - (N/2)}{N/2},\tag{2.4}$$

where n_e is the number of experts who deem the item as essential and N is the total number of experts. For example, we may construct an instrument that asks parents about aggression in their children. One item might ask, "Does your child bite other children?" If we ask 20 experts if they think this item is essential to measuring aggression in children and 17 agree it is, then the CVR can be calculated using the cvr function in **hemp**.

```
cvr(N = 20, n_e = 17)
```

```
[1] 0.7
```

We find that the CVR is 0.70 for this particular item, but we do not know whether 0.70 is large enough to keep the item in the instrument. Table 1 in Lawshe (1975) provides threshold values of CVR given a set number of experts. For 20 experts, the CVR minimum is 0.42 and we would conclude that experts think this item is essential beyond chance alone, and likely retain this item in our instrument.

Other forms of validity evidence measure the extent to which test scores relate to some external criterion (criterion-related validity evidence). Statistical support for this validity evidence may involve calculating simple correlations or the use of regression. Returning to the **interest** data set, we would expect that the vocabulary test (vocab) would be correlated with assessments that measure reading comprehension (reading) and sentence completion (sentcomp). Thus, we can use the cor function to calculate the Pearson correlation between these variables.

```
cor(interest[, c("vocab", "reading", "sentcomp")])
```

```
           vocab    reading  sentcomp
vocab    1.0000000 0.8030912 0.8132765
reading  0.8030912 1.0000000 0.7252155
sentcomp 0.8132765 0.7252155 1.0000000
```

The Pearson correlation between vocab and reading is 0.803, while the correlation between vocab and sentcomp is 0.813. This would represent concurrent validity evidence if the vocabulary tests were administered at the time of the reading and sentence completion assessments. If the vocabulary test predates the reading and sentence completion tests, then the Pearson correlation represents evidence supporting predictive validity.

Assume that we are measuring someone's interest in becoming a teacher (teacher) using a personality measure of social dominance (socdom) and we want to understand if there is additional predictive ability of administering the reading comprehension assessment beyond the personality measure alone. We are trying to assess whether reading comprehension has incremental validity, and that can be assessed within a regression framework using stepwise regression.

We do this by fitting two linear regression models. The first model (`mod_old`) regresses teacher on socdom, while the second model (`mod_new`) regresses teacher on both socdom and reading. A linear regression model can be estimated using the `lm` function in R. This function requires specifying a regression formula with a dependent variable (teacher) predicted by one or more independent variables (socdom and reading), and the name of the data set that includes these variables (`interest`).

```
mod_old <- lm(teacher ~ socdom, interest)
mod_new <- lm(teacher ~ socdom + reading, interest)
```

The resulting models are nested because `mod_new` includes the same dependent and independent variables used in `mod_old` while also including the reading variable. To examine the contribution of reading beyond socdom in predicting teacher, we can extract the change in the R-squared (R^2) value between the two models. In addition, we can compare the models statistically using the `anova` function in R. This test examines whether the R^2 change between the two models is statistically significant (i.e., $R^2 > 0$). If the R^2 change is statistically significant, then we can conclude that reading explains a significant amount of variation in the dependent variable teacher beyond socdom.

```
summary(mod_new)$r.squared - summary(mod_old)$r.squared
anova(mod_old, mod_new)
```

```
[1] 0.09125979
Analysis of Variance Table

Model 1: teacher ~ socdom
Model 2: teacher ~ socdom + reading
  Res.Df    RSS Df Sum of Sq      F    Pr(>F)
1    248 244.98
2    247 221.03  1    23.951 26.765 4.754e-07 ***
```

```
---
Signif. codes:
0 '***' 0.001 '**' 0.01 '*' 0.05 '.' 0.1 ' ' 1
```

From the output above, we see that the reading comprehension assessment has incremental validity beyond the social dominance measure alone ($p <$ .001) and that it explains approximately 9% more variability in interest in the teaching profession.

2.4.3 Item Analysis

Item analysis plays an important role in developing and revising measurement instruments. Item analysis can be performed using the base functions that come with R. We return to the SAPA data set to demonstrate how to perform item analysis for dichotomously scored items.

A common statistic to calculate when performing an item analysis is the proportion of examinees correctly answering each item. This is known as item difficulty, p, or the p value. This is not to be confused with item difficulty in item response theory (see Chapter 5) or a p-value from statistical hypothesis testing. The item with the highest item difficulty, ironically, is the easiest item for the examinees, which is why item difficulty is sometimes referred to as item easiness.

We can use the `colMeans` function to calculate item difficulty. Because we have examinees that have missing responses on some of the items, we need to pass the `na.rm = TRUE` argument to ignore the missing values in the computation. Otherwise, the `colMeans` function would return "NA" for the items that have at least one missing value. To make the item difficulties easier to read, we round them to three decimals using the `round` function.

```
item_diff <- colMeans(SAPA, na.rm = TRUE)
round(item_diff, 3)
```

```
  reason.4 reason.16 reason.17 reason.19   letter.7 letter.33
     0.640     0.698     0.697     0.615      0.600     0.571
letter.34 letter.58 matrix.45 matrix.46 matrix.47 matrix.55
    0.613     0.444     0.526     0.550     0.614     0.374
 rotate.3  rotate.4  rotate.6  rotate.8
    0.194     0.213     0.299     0.185
```

The output shows that the items reason.16 and reason.17 had the highest item difficulties, while rotate.8 had the lowest item difficulty. Roughly 70% of the students were able to answer reason.16 and reason.17 correctly, while only 19% could answer rotate.8 correctly.

Another widely used statistic in item analysis is item discrimination, which refers to the discriminatory power of the item in distinguishing examinees with a high ability from those with a low ability. Although there are many ways of computing item discrimination, the most common form is the point-biserial correlation between examinees' responses to the item and their total score on the test. Large, positive values indicate a strong relationship between answering the item correctly and having a high score on the test, while values close to zero indicate no relationship, and negative values indicate that answering the item correctly is associated with a lower overall test score. Values that are close to zero or negative suggest that the item may not be functioning properly. Some of the reasons for obtaining low or negative item discrimination could be using a wrong answer key for the item, having multiple correct answers, or no correct answer at all. Regardless of the cause, items with low or negative point-biserial correlations need to be modified, if the test/instrument is in a revision state, or removed from the test and scoring.

To calculate item discrimination for SAPA, we first compute the total test score using the rowSums function along with the na.rm = TRUE option and save it as total_score. Next, we correlate the items in SAPA with the total test score using the cor function. We specify the use = "pairwise.complete.obs" argument in the cor function because of missing responses. Finally, we save the correlation matrix as item_discr and print it.

```
total_score <- rowSums(SAPA, na.rm = TRUE)
item_discr <- cor(SAPA, total_score,
                  use = "pairwise.complete.obs")
item_discr
```

```
                [,1]
reason.4   0.5875787
reason.16  0.5339592
reason.17  0.5859068
reason.19  0.5582773
letter.7   0.5846261
letter.33  0.5569431
letter.34  0.5946924
letter.58  0.5760896
matrix.45  0.5095047
matrix.46  0.5149072
matrix.47  0.5478686
matrix.55  0.4465107
rotate.3   0.5100778
rotate.4   0.5559848
rotate.6   0.5542336
rotate.8   0.4806727
```

The results show that all the items in SAPA are moderately, positively correlated with the total test score. This indicates that all the items are functioning properly and provides no salient information for which items to remove or modify.

Another way of calculating item discrimination is to split examinees into two groups (e.g., 1 = high achievers and 0 = low achievers) based on their total test scores, and correlate this grouping variable with item responses. This is known as the item discrimination index. One option to create high and low achieving groups is to select the upper 27% and lower 27% of the examinees based on their total test scores. Here it should be noted that the decision to use 27% is somewhat arbitrary. We could easily use another value (e.g., 10% or 20%) to define the high and low achieving groups. After we define the cut-off for our groups, we calculate the proportion of examinees that answered the item correctly in the high and low achiever groups.

In the following example, we calculate the item discrimination index for reason.4 in the **SAPA** data set using the `idi` function from the **hemp** package. To specify the high and low achieving groups, we use `perc_cut = .27` in the `idi` function.

```
idi(SAPA, SAPA$reason.4, perc_cut = .27)
```

```
Upper 27% Lower 27%
 0.805136  0.194864
```

We find that 81% of the examinees in the high achieving group got the item correct, while only 19% of the examinees in the low achieving group got the item correct. This suggests that the item was easier for the high achievers and difficult for the low achievers. Therefore, we can say that this particular item would be useful for differentiating between the two groups but not necessarily within each group.

In addition to item difficulty and item discrimination indices, another useful statistic for item analysis is the item-reliability index. The item-reliability index (IRI) is defined as:

$$IRI = S_i * r_{i,tt}, \qquad (2.5)$$

where S_i is the standard deviation of the item i and $r_{i,tt}$ is the correlation between item i and the total test score. IRI can theoretically range between -0.5 and 0.5, with large, positive values indicative of high reliability. Below we calculate the IRI for all the items in the **SAPA** data set. We can do this using the `iri` function in **hemp**.

```
iri(SAPA)
```

```
              [,1]
reason.4  0.2820989
reason.16 0.2451971
```

```
reason.17 0.2692675
reason.19 0.2717135
letter.7  0.2865325
letter.33 0.2757209
letter.34 0.2897118
letter.58 0.2863221
matrix.45 0.2544930
matrix.46 0.2562540
matrix.47 0.2668171
matrix.55 0.2161230
rotate.3  0.2016459
rotate.4  0.2276081
rotate.6  0.2539219
rotate.8  0.1867207
```

The results returned from the `iri` function show that the IRI ranges from approximately 0.19 to 0.29 for the `SAPA` data set. These are all reasonable values for IRI (i.e., none are negative or close to zero).

When an external criterion is used instead of the total test score, this index is known as the item-validity index (IVI). The IVI can also range between -0.5 and 0.5, with large values (in absolute magnitude) indicative of higher validity. Large negative values are indicative of higher validity when the items are expected to correlate negatively with the criterion.

In the following example, we use the `ivi` function in **hemp** with reason.17 as the external criterion and reason.4 as the item of interest and find the IVI to be 0.19.

```
ivi(item = SAPA$reason.4, crit = SAPA$reason.17)
```

```
[1] 0.1903219
```

Another important aspect of items that needs to be analyzed is the response options. In the context of multiple-choice testing, alternative (i.e., incorrect) response options are referred to as distractors. Distractors play an role in a multiple-choice item. To ensure high-quality multiple-choice items, it is crucial to include well-functioning, plausible distractors that are more likely to attract examinees with partial knowledge. Distractors that are not plausible may need to be rewritten or replaced with a better distractor. Distractor quality is typically evaluated using distractor analysis (Gierl, Bulut, Guo, & Zhang, 2017). Distractor analysis is often conducted by looking at the proportion of examinees choosing a particular distractor (see Gierl et al. (2017) for a comprehensive review of distractor analyses).

To demonstrate distractor analysis, we use the items from the `multiplechoice` data set in **hemp**. This is a hypothetical multiple-choice test consisting of 27 multiple-choice items administered to 496 examinees. The four response options were coded as 1, 2, 3, and 4 in the data set. The answer key for the items can be seen by running the ?multiplechoice command in R. We use the distract function in **hemp** to calculate the proportion of examinees selecting each distractor. We save the results of distractor analysis as distractors and then print the results for the first six items with the head function.

```
distractors <- distract(multiplechoice)
head(distractors)
```

```
          1     2     3     4
item1 0.044 0.058 0.052 0.845
item2 0.109 0.069 0.792 0.030
item3 0.188 0.562 0.058 0.192
item4 0.034 0.125 0.742 0.099
item5 0.351 0.254 0.042 0.353
item6 0.081 0.198 0.558 0.163
```

In the table above, we see that all the items had distractors that were selected approximately 5% of the time or less. These distractors might be candidates for revision as they were endorsed at such a low level as to suggest most examinees did not see them as viable, plausible options. For item 1, the distractors were all functioning roughly the same (i.e., approximately 5% of the time each one was endorsed), suggesting they were all functioning well with respect to one another but that the item was too easy (the correct answer was option 4, and was selected by 84.5% of the examinees). In contrast, item 5 was a more difficult item, with the correct answer again being option 4. Options 1 and 2 were very likely misconceptions, while option 3 might be revised or potentially dropped from this item because of the low endorsement rate (only 4.2%). Given the very high endorsement of option 1 (35.1%), it is very likely that this option was also correct. To analyze this item further, calculating the item discrimination index for this item would be very helpful, as it would shed additional insight into overall item functioning.

2.5 Summary

In this chapter, several foundational measurement topics were introduced. The chapter focused on scales of measurement, validity, reliability, the CTT framework, and item analysis. In addition, we introduced the bootstrap as an option to obtain confidence interval for parameters when their sampling distributions are unclear. Because of the overview nature of this chapter, readers may want to refer to a more traditional introductory measurement textbook to get more detailed information about the statistics and methods presented. We recommend Thorndike and Thorndike-Christ (2010) for a good, general introduction to measurement with a focus on education, and R. J. Cohen et al. (2013) for a more general approach to testing in psychology.

Many great packages exist for R that expand on the statistics presented in this chapter. For example, the **psych** package (Revelle, 2017) is a general-purpose psychometric package and will be covered in greater detail in Chapter 4. In the next chapter, we address some of the shortcomings within the CTT framework, and provide a more powerful and flexible model for calculating reliability.

3

Generalizability Theory

3.1 Chapter Overview

This chapter presents generalizability theory, which is also known as G theory in the literature. The chapter begins with a brief review of G theory and defines the key concepts in G theory, specifically the distinction between a generalizability study (G study) and a decision study (D study); facets and units of measurement; study design; and fixed versus random facets. Several examples, using the **hemp** and **lme4** (Bates, Mächler, Bolker, & Walker, 2015) packages, demonstrate how to run G and D studies using R for a variety of designs.

3.2 Introduction

In measurement, our main interest lies in estimating an individual's underlying true score on a construct or a latent variable. For example, we might be interested in estimating examinees' knowledge of algebra, their intelligence quotient, their executive functioning skills, or their quality of life. Unfortunately, all of these estimates, regardless of our best efforts, will be imperfect and contain a certain level of measurement error.

Various models have been proposed to adjust for these inescapable errors in the measurement process. In Chapter 2, the classical test theory (CTT) model was introduced, which is the simplest and most commonly applied model in practice. Recall that the CTT model postulates that an individual's observed score on a construct is a sum of their true score on that construct and error (the first line in Equation 3.1) and subsequently the observed variance in the test scores is the sum of the true score variance and the residual (i.e., error) variance (the second line in Equation 3.1, see Chapter 2 for details). Mathematically we can define the CTT model as

$$X = T + E, \text{ and}$$
$$\sigma_X^2 = \sigma_T^2 + \sigma_E^2. \tag{3.1}$$

In the CTT model, all sources of observed variance that cannot be attributed to true score variance are captured in a single residual term. Therefore, no attempt is made to differentiate and quantify the various sources of error. All sources of error contained in the error variance are treated in the same manner regardless of whether they are malleable, i.e., within the controls of the test developer/administrator, and whether they are not malleable.

In generalizability theory (G theory), our primary focus is on disentangling the error variance into sources of variation that are controllable (Shavelson & Webb, 1991). We want to be able to disentangle this variance into sources of variation that are associated with the conditions of measurement, known as facets, which are manipulable by practitioners and researchers, and with our units of measurement (typically students or study participants). By disaggregating the error into various sources, we can quantify how altering the levels of a facet(s) affects the reliability of our test scores, and we can adjust our instrument accordingly.

As a concrete example, imagine creating an instrument to measure graduate school readiness. Graduate programs train students to become researchers who are capable of presenting their ideas through research papers, manuscripts, technical reports, and other outlets. Therefore, it is necessary for applicants to have strong writing skills to produce well-reasoned and articulated essays when they apply for and enter graduate school. To measure the writing skills that are essential for graduate school, we construct an instrument consisting of a series of writing prompts to which applicants need to respond. The writing prompts that we have selected for our instrument might represent all the writing prompts that we care about (i.e., there are a fixed number of writing prompts) and we would say that the writing prompts are a fixed facet. This is analogous to a fixed effect using the ANOVA terminology. More likely, the writing prompts represent a random sample of an extremely large, potentially infinite pool of writing prompts that we could have selected from. In the context of ANOVA, writing prompts would be treated as a random effect and we would say the writing prompts are a random facet.

After we have constructed the instrument, we then administer the assessments to the examinees (i.e., prospective graduate students). The prospective students are given an hour to craft their responses to our writing prompts. Their responses are then scored by a rater. As with the writing prompts, we are generally not interested in any particular rater but all the raters, and thus we would consider the raters as a random facet. If the raters could conceivably score all the writing prompts, then writing prompts and raters would be considered crossed; however, if a rater could only score a certain subset of the writing prompts, then the raters would be considered nested within the writing prompts. In our example, we will assume that all raters could score any writing prompts and thus the raters and the writing prompts are crossed.

What we aim to achieve is an estimate of the students' true writing abilities across all the writing prompts and all the raters. That said, our aim in G theory is to obtain an estimate of the writing ability that can be considered generalizable and dependable across all the potential sources of variation (Brennan, 1992). In this hypothetical example, students, writing prompts, and raters all represent sources of variation, with the writing prompts and raters representing the facets and the students representing our unit of measurement. Variation associated with the students would be true score variance, or in G theory language, universe score variance.

We can decompose the observed score for a student (s) on a single writing prompt (w) by a single rater (r), X_{swr}, as:

$$X_{swr} = \mu + \nu_s + \nu_w + \nu_r + \nu_{sw} + \nu_{sr} + \nu_{wr} + \nu_{swr,e}. \qquad (3.2)$$

In Equation 3.2, we see that a student's score contains eight components: the grand mean (μ); a student effect (ν_s); a writing prompt effect (ν_w); a rater effect (ν_r); a student by writing prompt effect (ν_{sw}); a student by rater effect (ν_{sr}); a writing prompt by rater effect (ν_{wr}); and finally, a student by writing prompt by rater effect, which is completely confounded with the residual term ($\nu_{swr,e}$). Each of these effects, because we are assuming they are random effects (random variables), have a distribution, which we usually assume is a normal distribution, with a mean of 0 and some unknown variance (except for the grand mean, which is a constant). Furthermore, we can express the total variance in the observed scores on the writing prompts over the entire universe (i.e., over all possible items and raters) and the population (i.e., over all students) for this design as:

$$\sigma^2(X_{swr}) = \sigma_s^2 + \sigma_w^2 + \sigma_r^2 + \sigma_{sw}^2 + \sigma_{sr}^2 + \sigma_{wr}^2 + \sigma_{swr,e}^2. \qquad (3.3)$$

In Equation 3.3, we see that the variance in the observed scores can be written as seven independent variances: The variance in students' writing abilities, which in G theory is known as universe-score variance, is the analogous to CTT's true score variance (σ_s^2); the variation in writing prompts (σ_s^w, e.g., the extent to which some prompts might be easier to respond to than others); the variation in raters (σ_r^2, the extent to which some raters give higher or lower scores than others); the variance in student-by-writing prompt (σ_{sw}^2, the extent to which some writing prompts might be more easily answered by some students than others); the student-by-rater variance (σ_{sr}^2, the extent to which a rater scores certain students higher than others); the writing prompt-by-rater variance (σ_{wr}^2, the extent to which a rater scores some prompts higher than others); and finally, the student-by-writing prompt-by-rater variance ($\sigma_{swr,e}^2$, the extent to which a student receives a higher score on a certain prompt by a certain rater) is completely confounded with the residual variance. In CTT, the variances, $\sigma_w^2 \ldots \sigma_{swr,e}^2$, would all be captured by the single residual variance, σ_E^2.

Figure 3.1 shows a Venn diagram of the hypothetical design outlined above. In Figure 3.1, the sources of variability associated with students, writing prompts, and raters and their interactions are shown. The amount of overlap between the circles (the sources of variation) represents the magnitude of the interactions.

At this point, the advantage of G theory over CTT becomes more apparent. By decomposing total variation into manipulatable facets, we are able to quantify how much variation is associated with each condition of measurement and can alter the number of levels associated with each of these conditions in order to obtain a certain level of reliability.

In G theory, two types of studies are usually conducted: a generalizability (G) study and a decision (D) study (Shavelson & Webb, 1991). In a G study, we identify all the potential sources of variation, incorporate them into our study design, and thus define our universe of admissible observations. In our writing prompt example, any rater could evaluate any writing prompt, so our universe of admissible observations would be that any combination of a prompt and rater is admissible. In a G study, we estimate the variances shown in Equation 3.3 and Figure 3.1.

In a D study, we investigate hypothetical scenarios to obtain generalizable and dependable scores. In our writing prompt example, we could assume that it is possible to alter the number of prompts, the number of raters, or both. What primarily guides the G and D studies is how we plan to use the test scores. If we wish to use the test scores to make relative decisions, such as decisions regarding the relative standing of our prospective graduate students, then we summarize our scores with the generalizability coefficient, which is the G theory equivalent to the CTT's reliability. If we want to use our test scores to make absolute decisions, such as whether a student is ready to write at a graduate school level, then we summarize our scores using the dependability coefficient. To find the generalizability coefficient, we need to calculate relative error variance, σ_{rel}^2; to find the dependability coefficient, we need to calculate the absolute error variance, σ_{abs}^2. The generalizability and dependability coefficients are then calculated as:

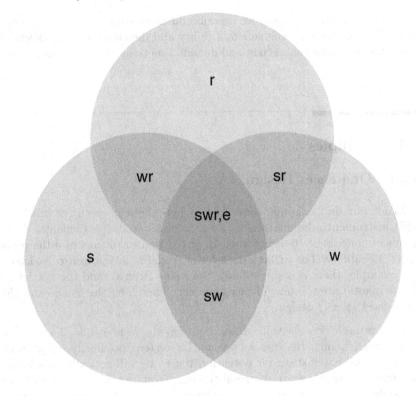

FIGURE 3.1
Venn diagram of a two-facet crossed (p x w x r) design. Note that s stands for student (the unit of measurement), w stands for writing prompt, r stands for rater, and e stands for error.

$$\text{generalizability coefficient} = \rho^2 = \frac{\sigma_s^2}{\sigma_s^2 + \sigma_{rel}^2}, \text{ and} \quad (3.4)$$

$$\text{dependability coefficient} = \phi = \frac{\sigma_s^2}{\sigma_s^2 + \sigma_{abs}^2}, \quad (3.5)$$

where σ_s^2 is the variation of students' test scores (the universe-score variance). What goes into σ_{rel}^2 and σ_{abs}^2 depends on the design of the study. For our readers who might be interested in knowing more about what goes into these variance components and, more generally, about G and D studies, we strongly recommend *Generalizability Theory: A Primer* by Shavelson and Webb (1991). This primer is easy to follow, walks through a variety of G study designs, and can help the interested reader decide on an appropriate study design.

In the following sections, we describe different study designs and demonstrate how to calculate the generalizability and dependability coefficients for these designs using the `gstudy` and `dstudy` functions in the **hemp** package.

3.3   Examples

3.3.1   One-Facet Design

Data for our first example come from a hypothetical executive functioning (EF) instrument administered to a group of 30 study participants. The instrument consists of 10 dichotomously scored items that measure the participants' EF ability. The `efData` data set is available in the **hemp** package. In this example, there is a single facet, the items (Items), and the participants (Participants) are the unit of measurement. Therefore, this is an example of a one-facet (p x i) design.

The variable Score within `efData` contains the responses to the item given by the participants. Because we are not interested specifically in the administered items but instead all potential items, we will consider this facet as random. In general, we recommend treating facets as random unless there is a specific reason to treat a facet as fixed. Figure 3.2 shows the sources of variation in this design, and the observed variation in the total scores on the executive function can be written as:

$$\sigma^2(X_{pi}) = \sigma_p^2 + \sigma_i^2 + \sigma_{pi,e}^2. \tag{3.6}$$

In Figure 3.2 and Equation 3.6, we see that there are three sources of variation in the study: variation associated with participants (p), variation associated with the items (i), and variation associated with the interaction of items and participants (pi,e). In a one-facet design, this interaction is completely confounded with the random error (Shavelson & Webb, 1991).

3.3.1.1   G Study

In the G study, we estimate the variance components that are defined in Equation 3.6. To estimate these variance components, first the **hemp** package is loaded using the `library` command.

```
library("hemp")
```

The **hemp** package uses the `lmer` function in the **lme4** package to esti-

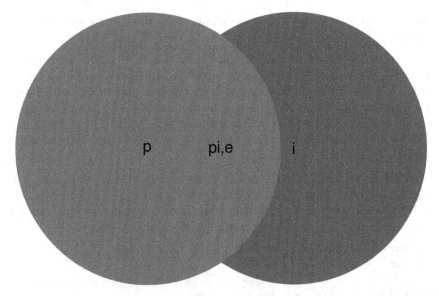

FIGURE 3.2
Venn diagram of a one-facet (p x i) design. Note that p stands for participant (the unit of measurement), i stands for item, and e stands for error.

mate the variance components. To estimate the participant and item variances, we need to use **lme4**'s syntax for specifying random effects. The **lme4** syntax for these effects are (1 | Participants) and (1 | Items), respectively. Within the same syntax, Score serves as the dependent variable that we are trying to decompose. After we fit the model in **lme4** and save the results as onefacet_mod, we call the gstudy function on the model (i.e., onefacet_mod). We save the G study as onefacet and then print the results at the end.

```
onefacet_mod <- lmer(Score ~ (1 | Participants) + (1 | Items),
                     data = efData)
onefacet <- gstudy(onefacet_mod)
onefacet
```

```
       Source Est.Variance Percent.Variance
1 Participants       0.0258             9.9%
2        Items       0.0959            36.8%
3     Residual       0.1387            53.3%
```

The estimated variances from the G study reflect the magnitude of error in generalizing from a participant's score on a single EF item to their universe score. These are not the variances associated with generalizing based on our present test length of 10 items. If we want to know this, which we undoubtedly do, we need to consider a test consisting of 10 items when we conduct a D study.

A useful heuristic for interpreting the variances that are estimated in a G study is to calculate the percent of the total variance that each variance component represents.[1] These percentages are presented in the last column of the printed data frame above. For a one-item test, we find that the variance component for participants (i.e., the universe score variance) accounts for only 10% of all the variance. This is rather low, suggesting that, unsurprisingly, a one-item test is not very reliable.

To help with the interpretation of the universe score component, a histogram can be created. The R code presented below loads the **lattice** package, calculates the percent of items that each participant got correct by using the **aggregate** function, and then creates a histogram (see Figure 3.3).

```
library("lattice")
scores <- aggregate(efData$Score,
                 by = list(efData$Participants), mean)
histogram(~ x, scores, type = "count",
          xlab = "Proportion of Items Correct")
```

The histogram shows that none of the participants got all the items correct or incorrect and that the overwhelming majority of participants got 50% or 60% of the items correct on the test (i.e., 5 or 6 correct answers). Because of this tight clustering, the universe score variance was low.

The variance component for the items (0.0959, or 36.8% of the total variance) is large relative to the universe score variance but smaller than the residual variance. Shavelson and Webb (1991) suggest taking the square root of this variance to obtain the standard deviation and then calculating the range for 4 standard deviations (roughly 95% of the scores if one assumes a normal distribution). The standard deviation is approximately 0.31 and multiplying this by four is approximately 1.2. Given that the item means can only range from 0 to 1, a standard deviation of 0.31 suggests that there is a lot of variation in the means of the items.

Next, we calculate the mean for each item again using the **aggregate** function and save the item means as **item_means**. Before we print the results, we rename the columns of this new data set as "Item" and "Mean" using the **colnames** function.

```
item_means <- aggregate(efData$Score,
                 by = list(efData$Items), mean)
colnames(item_means) <- c("Item", "Mean")
item_means
```

---

[1]This is also known as intraclass correlations.

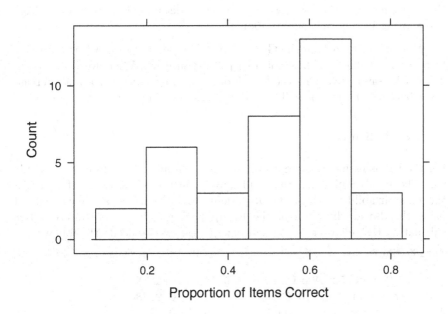

FIGURE 3.3
Histogram of the percent of items correct on the EF instrument.

	Item	Mean
1	1	0.94285714
2	2	0.68571429
3	3	0.68571429
4	4	0.08571429
5	5	0.74285714
6	6	0.77142857
7	7	0.08571429
8	8	0.60000000
9	9	0.45714286
10	10	0.11428571

The item means range from .09 to 94, which is quite a lot of variation and corroborates the high percent of variation accounted for by the items.

The large residual variance captures both the participant by item interaction and the random error (which we are unable to disentangle). Maybe some items were more easily answered by some participants or maybe there was systematic variation (e.g., the physical environment where the assessment was administered or the temperature in the room), or possibly just random variation (e.g., some participants were hungry or tired when they completed the assessment). Either way, we are unable to disentangle these sources from one another in this variance component.

We can conclude from the G study that a one-item test of EF is not going to be very reliable, which is not a surprising outcome. What we would likely do next is consider varying the length of our test to reduce error and increase the reliability of our scores. This can be examined in a D study.

3.3.1.2   D Study

For the EF assessment, there is only one facet (items). Therefore, we can only vary the number of items that will appear on the EF assessment to affect the generalizability or dependability coefficients. D studies can be carried out using the `dstudy` function in the **hemp** package. If we were interested in calculating the reliability of a one-item EF test, we would do the following:

```
dstudy(onefacet, unit = "Participants", n = c("Items" = 1))
```

```
            Source Est.Variance  N Ratio of Var:N
1 Participants        0.0258 350         0.0258
2        Items        0.0959   1         0.0959
3     Residual        0.1387   1         0.1387
```

```
The generalizability coefficient is: 0.1568389.
The dependability coefficient is: 0.09907834.
```

Three pieces of information are printed by default when we call the `dstudy` function in the **hemp** package:

1. a data frame containing the sources of variance, the estimated variance, the number of levels of the participants and facets, and this variance divided by the number of levels of the facet,

2. the generalizability coefficient, and

3. the dependability coefficient.

The D study shows that a test consisting of only one item has a generalizability coefficient of approximately .16 and a dependability coefficient of .10. Given our current test length of 10, we would be more interested in knowing the reliability for a test length of 10 items. We change the "n" argument to n = c("Items" = 10) in the `dstudy` function. We save the results as `onefacet_dstudy`.

```
onefacet_dstudy <- dstudy(onefacet,
                   unit = "Participants",
```

```
                     n = c("Items" = 10))
onefacet_dstudy
```

```
         Source Est.Variance   N Ratio of Var:N
1 Participants        0.0258 350        0.02580
2        Items        0.0959  10        0.00959
3     Residual        0.1387  10        0.01387
```

```
The generalizability coefficient is: 0.6503655.
The dependability coefficient is: 0.5237515.
```

With our current test length, the generalizability coefficient is .65 and the dependability coefficient is .52. These reliabilities are still quite low, suggesting that we would want to increase the test length even more to improve the reliability of the scores.

Remember that we saved the output from the dstudy function as onefacet_dstudy. Using the saved results in onefacet_dstudy, we can extract the relative error variance (relvar) and the absolute error variance (absvar).

```
onefacet_dstudy$relvar
```

```
[1] 0.01387
```

```
onefacet_dstudy$absvar
```

```
[1] 0.02346
```

An effective way to visualize the effect of increasing test length on reliability is to plot the coefficients against the test length. This can be done using the dstudy_plot by specifying a range of values for test length that we want the reliability to be calculated for. In the call below, the dependability coefficient (specified via the g_coef = FALSE argument) is plotted against different test lengths (i.e., 10 items to 60 items).

```
dstudy_plot(onefacet, unit = "Participants",
            facets = list(Items = c(10, 20, 30, 40, 50, 60)),
            g_coef = FALSE)
```

In Figure 3.4, we see that doubling the length of our current test, from 10 to 20 items, results in the largest gains in reliability. Depending on the use of the test scores (e.g., are they high stakes or low stakes absolute decisions), we may be satisfied with a reliability of just under .80 for a test length of 40 items, or if higher reliability is necessary (e.g., a reliability of .90), we may want to have a test that is six times longer than the current test (i.e., 60 items in length). The utility of plotting various D studies against a reliability

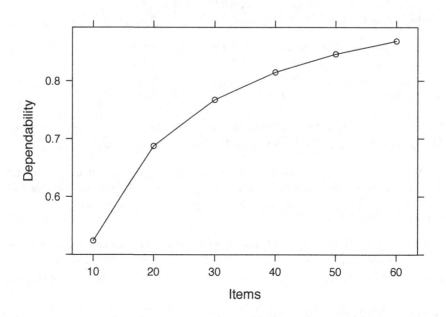

FIGURE 3.4
Plot of dependability coefficient against number of EF items for the one-facet
random design.

coefficient is that it allows us to quickly assess the impact of increasing the
number of levels of a facet on reliability.

3.3.2   Two-Facet Crossed Design

To demonstrate a two-facet crossed design, we will use the `writing` data set in
the **hemp** package. This hypothetical data set contains data on 10 students
(s) measured on 5 writing prompts (w) scored by two raters (r). The data
are in long format where each row corresponds to a single observation (i.e., a
score on a single writing prompt), and the variables student, prompt, rater,
and score contain the student identifier, the writing prompt identifier, the
rater identifier, and the score for a single observation.

3.3.2.1   G Study

Equation 3.3 and Figure 3.1, shown above, present the variances that are estimated in this two-facet crossed (p x w x r) design. Variances for student,

writing prompt, rater, and their interactions are estimated. Unlike in the one-facet design above, we must specify an interaction in our call to **lme4** because all two-way interactions can be estimated. However, as indicated above, the student-by-writing prompt-by-rater interaction is confounded within the residual variance and is not specified. To indicate an interaction, **lme4** uses the syntax (1 | variable1:variable2). This syntax is similar to how interactions are specified in a call to **lm**. In the call to the **lmer** function below, we specify six variances (i.e., students, prompts, raters, students-by-prompts, students-by-raters, and prompts-by-raters).

```
twofacet_mod <- lmer(scores ~ (1 | students) + (1 | prompts) +
                     (1 | raters) + (1 | students:prompts) +
                     (1 | students:raters) +
                     (1 | prompts:raters),
                     data = writing)
```

The `twofacet_mod` object is then passed to the `gstudy` function so that the variances can be estimated for the G study.

```
twofacet <- gstudy(twofacet_mod)
twofacet
```

	Source	Est.Variance	Percent.Variance
1	students:prompts	12.8883	3.6%
2	students:raters	62.4073	17.4%
3	prompts:raters	4.8772	1.4%
4	students	180.1760	50.2%
5	prompts	8.1517	2.3%
6	raters	19.8028	5.5%
7	Residual	70.3428	19.6%

In the output above, the interactions are printed first followed by the variance for students, writing prompts, raters, and finally the residual variance. The largest variance component, the variance for students (180.176), accounts for about 50% of the total variance in scores. This is the variance component for the universe scores and indicates that students systematically differed in their scores on the writing assessment. The variance component for raters is small relative to students but larger than the writing prompt variance. If we take the square root of this variance ($\sqrt{19.8028} = 4.45$), the expected range of rater means is about 18 or 27 (assuming a normal distribution and 2 or 3 standard deviations). This range is reasonably small relative to the range in scores on a writing prompt (0 to 100). Similarly, the variance of writing prompts was even smaller and the expected range in scores on the writing prompts is quite small (12 or 18). The student-by-writing prompt and the writing prompt-by-rater variances were quite small, indicating that students' ranks did not differ much depending on the writing prompts and that raters scored the items reasonably consistently. The student-by-rater variance (62.4073) is quite large relative to the other variance components. This finding suggests that some raters scored some students higher than other students. Finally, the residual variance (70.3428 or about 20% of all variance) shows a substantial proportion of the variance was due to the three-way interaction between students, raters, and writing prompts and/or other random or systematic sources of variation that were not accounted for in the study.

3.3.2.2   D Study

The generalizability and dependability coefficients for this two-facet crossed design with 2 raters and 5 writing prompts can be calculated using the dstudy function.

```
dstudy(twofacet, n = c("raters" = 2, "prompts" = 5),
       unit = "students")
```

```
              Source Est.Variance   N Ratio of Var:N
1 students:prompts       12.8883    5        2.57766
2  students:raters       62.4073    2       31.20365
3   prompts:raters        4.8772   10        0.48772
4         students      180.1760  100      180.17600
5          prompts        8.1517    5        1.63034
6           raters       19.8028    2        9.90140
7         Residual       70.3428   10        7.03428

The generalizability coefficient is: 0.815307.
The dependability coefficient is: 0.7732509.
```

The generalizability coefficient is 0.82 and the dependability coefficient is 0.77. For a low stakes exam, these coefficients are at an acceptable level. However, reliability greater than 0.90 would be desirable for a high stakes exam.

Assuming our writing assessment is going to be used for making high stakes decisions, we can use the dstudy_plot function to determine what kind of a design would give us a generalizability coefficient of 0.90 or higher. Below we plot the generalizability coefficient against the number of writing prompts (ranging from 3 to 8) by the number of raters (ranging from 1 to 5).

```
dstudy_plot(twofacet, unit = "students",
            facets = list(prompts = 3:8,
                          raters = 1:5))
```

Figure 3.5 shows the expected generalizability coefficients by writing prompts and raters. If we are interested in obtaining a reliability of 0.90, we need to have at least four raters. If we have four raters, then we need approximately seven writing prompts and if we have five raters, we need about

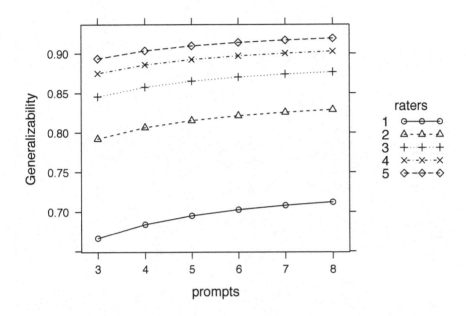

FIGURE 3.5
Plot of generalizability coefficient against number of writing prompts by number of raters.

four writing prompts. The decision to use one design over another may depend on availability of raters or the number of prompts that we wish to make our test takers respond to. We may also decide whether the additional gain in reliability by having more than three writing prompts and two raters is worth it.

3.3.3 Two-Facet Partially Nested Design

To demonstrate a two-facet, partially nested design, we use the `writing2` data set in the **hemp** package. This hypothetical data set contains data on 10 students (s), measured on 5 writing prompts (w), scored by two raters (r). Unlike the `writing` data set where the raters scored all the writing prompts for all the students, in the `writing2` data set each rater scored only one student (i.e., raters were nested within students). For every student, two raters scored a student on all 5 writing prompts. However, each rater scored no more than one student. This design can be expressed as *r:s* x *w*, indicating that raters are nested within students and are crossed with writing prompts.

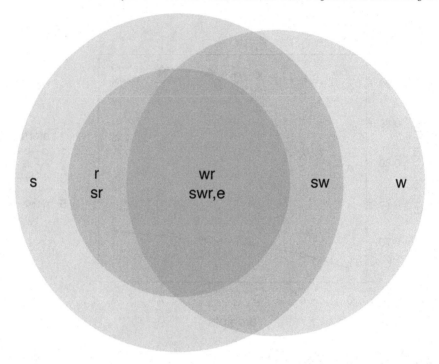

FIGURE 3.6
Venn diagram of a two-facet, partially nested design, $(r{:}s) \times w$. Note that s stands for student (the unit of measurement), w stands for writing prompt, r stands for rater, and e stands for error.

3.3.3.1   G Study

As with the `writing` data set, the `writing2` data are also in long format where each row corresponds to a single observation (i.e., a score on a single writing prompt). The variables student, prompt, rater, and score contain the student identifier, the writing prompt identifier, the rater identifier, and the score for a single observation.

Figure 3.6 shows a Venn diagram of the sources of variation for this design. In the partially nested design, the variation associated with raters is completely nested within subjects and the rater variation (r) is completely confounded with the subject-by-rater variation (sr). Figure 3.6 also shows that the writing prompt-by-rater variation (wr) is completely confounded with the student-by-writing prompt-byrater variation (swr) and the residual variance (e). The writing prompt variation (w) and the student-by-writing prompt variation (wr) can be estimated.

Fitting a nested design using `lmer` is similar to the crossed design; however, we need to specify that raters are nested within students. The syntax for specifying that raters are nested within students is (1 | students/raters). This will allow us to estimate the student variation and the variation associated with raters and students-by-raters, which are confounded. In the call to the lmer function, we must also specify the writing prompt variation, (1 | prompts), and the student-by-writing prompt variation, (1 | students:prompts). We save the results from this nested model as nested_model and then use it in the gstudy function.

```
nested_model <- lmer(scores ~ (1 | students/raters) +
                     (1 | prompts) +
                     (1 | students:prompts),
                     data = writing2)
nested <- gstudy(nested_model)
nested
              Source Est.Variance Percent.Variance
1 students:prompts      10.4497                 3%
2  raters:students      82.2100              23.6%
3          students     170.2747             48.8%
4           prompts      10.5903                 3%
5          Residual      75.2200             21.6%
```

The G study output prints the student-by-writing prompt variance (students:prompt); the student-by-raters variance (raters:students); the student (i.e., universe-score) variance (students); the writing prompt variance (prompts); and finally the residual variance (Residual), which also contains the student-by-writing prompt-by-rater variance and the writing prompt-by-rater variance.

The largest source of variation is students, accounting for nearly 50% of the total variation, followed by the variation associated with the student by raters variation (24%), and finally the residual variation and the other confounded sources of variation.

3.3.3.2 D Study

A nested design is specified in exactly the same manner as a crossed design with the dstudy function. A D study with 2 raters and 5 prompts is specified below.

```
dstudy(nested, n = c("raters" = 2, "prompts" = 5),
       unit = "students")

              Source Est.Variance   N Ratio of Var:N
1 students:prompts      10.4497     5        2.08994
2  raters:students      82.2100     2       41.10500
3          students     170.2747   100      170.27470
```

| 4 | prompts | 10.5903 | 5 | 2.11806 |
| 5 | Residual | 75.2200 | 10 | 7.52200 |

The generalizability coefficient is: 0.7705029.
The dependability coefficient is: 0.7631882.

  With this design, we have reasonably high reliability to make relative decisions (0.77) and absolute decisions (0.76). If we want to compare multiple designs, again we can use the `dstudy_plot` function.

3.3.4 Two-Facet Crossed Design with a Fixed Facet

As a final example, we again consider the `writing` data set. However, this time we will treat writing prompts as a fixed facet. In other words, we are not interested in generalizing over writing prompts but rather we want to know how different raters would score the same writing prompts. This represents a restriction of our universe of generalization and will affect the variances that are estimated and subsequently our reliability estimates. Since we are treating raters as random and writing prompts as fixed, the resulting model becomes a mixed model in the ANOVA framework.

3.3.4.1 G Study

In an earlier example, we fit a two-facet crossed design to the `writing` data set and saved it as `twofacet_mod`. This time we need to tell the `gstudy` function that we want to treat writing prompts as a fixed facet. To achieve this, we include `fixed = "prompts"` in the `gstudy` function and save the results as `twofacet_fixed`.

```
twofacet_fixed <- gstudy(twofacet_mod, fixed = "prompts")
twofacet_fixed
```

	Source	Est.Variance	Percent.Variance
1	raters	20.7782	7.4%
2	students	182.7537	65.3%
3	Residual	76.4758	27.3%

  The output contains fewer sources of variation than the variation for a fully crossed two-facet design because the sources of variation associated with writing prompts are subsumed within these sources of variations. Although it is not shown with writing prompts being a fixed facet, the student-by-writing prompt variation contributes to the student variation (universe score variance) and that is why this variance component is larger than the one in a two-facet design where both facets were random. Because of this, reliability will be greater in this particular design.

3.3.4.2   D Study

To show that a design with a fixed facet tends to have a higher reliability, we fit the same D study that we used earlier with 2 raters and 5 writing prompts.

```
dstudy(twofacet_fixed, n = c("raters" = 2, "prompts" = 5),
       unit = "students")
```

```
       Source Est.Variance   N Ratio of Var:N
1       raters      20.7782   2        10.38910
2     students     182.7537 100       182.75370
3     Residual      76.4758  10          7.64758

The generalizability coefficient is: 0.9598344.
The dependability coefficient is: 0.9101716.
```

The generalizability coefficient is 0.96 when writing prompts are fixed as compared to 0.82 when it was treated as random. This highlights the importance of considering whether a facet is random or fixed and the consequence of this decision on reliability.

3.4   Summary

In this chapter we introduced G theory. We presented various designs and how to run G and D studies for these particular designs. For readers who want to use more complex designs in G theory, we recommend the **gtheory** package (Moore, 2016). The **gtheory** package is capable of fitting both univariate and multivariate model, while the G theory functions in **hemp** can only fit univariate models. In the next chapter, we turn our attention to factor analysis. Factor analysis represents an important transition in this textbook. The models presented so far have focused on (observed) test scores, while factor analysis focuses on the individual items. Factor analysis also forms the foundation of item response theory, which we will introduce in Chapter 5.

4

Factor Analytic Approach in Measurement

4.1 Chapter Overview

Chapter 4 introduces the factor analytic approach in measurement. Factor analysis represents the link between the classical test theory presented in Chapter 3 and item response theory presented in Chapters 5 through 8. This chapter begins with a brief review of the common factor model. Exploratory factor analysis is then presented with examples using `factanal` and the `fa` function from the **psych** package. Within exploratory factor analysis, rotation, dimensionality, and factor score extraction are discussed. The chapter concludes with the confirmatory factor model and examples using the **lavaan** package. Examples using ratio/interval scale indicators and ordinal indicators are presented for both exploratory and confirmatory factor analyses.

4.2 Introduction

Factor analysis is a statistical modeling technique that aims to explain the common variability among a set of manifest variables or indicators with a reduced set of variables known as factors or dimensions. Let p be equal to the number of indicators and q be equal to the number of factors, then the common factor model can be expressed as

$$\mathbf{x} = \boldsymbol{\mu} + \boldsymbol{\Lambda}\mathbf{f} + \boldsymbol{\epsilon}, \qquad (4.1)$$

where $\mathbf{x}$ is a p-dimensional vector of continuous observed variables; $\boldsymbol{\mu}$ is a p-dimensional vector of scalars (i.e., intercepts); $\boldsymbol{\Lambda}$ is a $p \times q$ matrix of factor loadings; $\mathbf{f}$ is a q-dimensional vector of latent variables/factors where it is typically assumed that $E(\mathbf{f}) = \mathbf{0}$ and $Var(\mathbf{f}) = \boldsymbol{\phi}$; and $\boldsymbol{\epsilon}$ is a p-dimensional vector of random errors, i.e., unique or specific factors, where $Var(\boldsymbol{\epsilon}) = \boldsymbol{\psi}$.

If we let the true covariance matrix of $\mathbf{x}$ in our population be $\boldsymbol{\Sigma}$, then our goal in factor analysis is to find a parametric model, $\mathbf{\Sigma}(\boldsymbol{\theta})$, that describes $\mathbf{\Sigma}$ as closely as possible. Equation 4.1 may be re-expressed to reflect this purpose

$$Var(\mathbf{x}) = \mathbf{\Lambda}\boldsymbol{\phi}\mathbf{\Lambda}^T + \boldsymbol{\psi} = \mathbf{\Sigma}(\boldsymbol{\theta}), \tag{4.2}$$

where $\boldsymbol{\theta} = (\mathbf{\Lambda}, \boldsymbol{\phi}, \boldsymbol{\psi})$. Since we can never observe $\mathbf{\Sigma}(\boldsymbol{\theta})$, we must estimate $\mathbf{\Sigma}(\boldsymbol{\theta})$ with $\mathbf{\Sigma}(\hat{\boldsymbol{\theta}})$ subject to the constraint that $\mathbf{\Sigma}(\hat{\boldsymbol{\theta}})$ should be as close as possible to our sample covariance matrix. The types of constraints we place on the model in Equation 4.2 determine whether we are doing an exploratory factor analysis (EFA): indicators load on all factors and indicator residuals are uncorrelated or confirmatory factor analysis (CFA): indicators load only on hypothesized factors and indicator residuals are possibly correlated. Within an EFA framework, we must also specify whether we are using an orthogonal ($\boldsymbol{\phi}$ is a diagonal matrix, factors are uncorrelated) or oblique rotation ($\boldsymbol{\phi}$ is not a diagonal matrix, factors are correlated).

At the core of factor analysis is the desire to reduce the dimensionality of the data from p indicators to q factors, such that $q < p$. In this sense, factor analysis is similar to principal components analysis (PCA) but differs as factor analysis is a statistical model, whereas PCA is strictly a dimension-reduction technique. Factor analysis is also typically used when there is interest in assessing the feasibility of a theoretical relationship between the latent variables (i.e., factors). For readers who are interested in a more in-depth understanding of the theoretical and statistical framework underlying the common factor model, we strongly recommend Gorsuch (1983), Bollen (1989), Mulaik (2009), and Kline (2015).

4.3　Exploratory Factor Analysis (EFA)

During instrument development or when there are no prior beliefs about the dimensionality or structure of an existing instrument, an EFA should be considered to investigate the factorial structure of the instrument. When conducting EFA, no restrictions are placed on the common factor model presented in Equations 4.1 and 4.2. The major considerations within EFA are the number of factors to extract; whether more than one factor needs to be extracted; whether to do an orthogonal or oblique rotation; and which type of statistical estimation technique should be used depending on the nature of the data (e.g., categorical vs. continuous variables, normal vs. non-normal data).

The **stats** package (R Core Team, 2017) provides a capable, albeit relatively barebones, factor analysis function, `factanal`. The `factanal` function is minimal in that it is only capable of performing factor analysis with maximum likelihood. For many cases, maximum likelihood-based factor analysis will be sufficient. However, maximum likelihood has an assumption that the indicators are multivariate normal. Tests of multivariate normality available are for R, e.g., in the **MVN** package (Korkmaz, Goksuluk, & Zararsiz, 2014). However, we recommend visually inspecting the data before conducting a factor analysis. A common method for investigating multivariate normality in factor analysis is to assess the marginal normality of the manifest variables and transform variables as necessary. This method, however, does not assure multivariate normality, as univariate normality does not imply multivariate normality, while the converse does.

After presenting EFA based on maximum likelihood, we present an alternative statistical discrepancy function that uses ordinary least squares (OLS) estimation to find the minimum residual solution, using the **psych** package. The minimum residual solution is preferable in situations where maximum likelihood will not converge (e.g., singular matrices). However, it has its own share of quirks (e.g., occasionally having commonalities greater than 1).

4.3.1   EFA of a Cognitive Inventory

To demonstrate how to conduct an EFA in R, we will use the `interest` data set in the **hemp** package. The `interest` data set comes from a fictitious instrument given to 250 participants. The items in the instrument measure cognition, personality, and vocational interest. In the `interest` data set, the first three variables correspond to gender, years of education, and age. The next six variables are assumed to measure cognition, which we will focus on in the following example (for more information on the `interest` data set, `?interest` can be run in R).

To begin, we load the **hemp** package using the `library` command. Next, we take a subset of the `interest` data set that only includes the cognition items, and save this new data set as `cognition`. The `cognition` data set includes measures of vocabulary (vocab), reading comprehension (reading), sentence completion (sentcomp), mathematics (mathmtcs), geometry (geometry), and analytical reasoning (analyrea).

```
library("hemp")
cognition <- subset(interest, select = vocab:analyrea)
```

As with any statistical analysis, we should explore our data prior to fitting any factor analytic models. We perform this preliminary analysis for several reasons. We want to verify that the data have been correctly read into R, a moot exercise in this case as the data set were not read in; to understand the extent of missing data; to understand the marginal distributions of our indicators, the relationships between the indicators jointly, and whether certain observations are extreme and likely to be influential cases in our analyses. This preliminary analysis typically involves examining descriptive statistics and creating univariate and bivariate plots.

As mentioned in Chapter 1, we can use the summary function to extract descriptive data from a data frame. We supplement the descriptive statistics returned from the summary function by using the apply function to calculate the variances of the indicators. In the apply function, we specify the data set we are using (cognition), how the desired function should be applied (1 to all the rows, 2 to all the columns), and the name of the desired function we want to apply (var for variance).

```
summary(cognition)
```

```
     vocab                reading              sentcomp
 Min.   :-2.62000    Min.   :-2.4700    Min.   :-2.47000
 1st Qu.:-0.60500    1st Qu.:-0.5175    1st Qu.:-0.55000
 Median : 0.04000    Median : 0.1850    Median : 0.10500
 Mean   : 0.09016    Mean   : 0.1350    Mean   : 0.07356
 3rd Qu.: 0.86000    3rd Qu.: 0.7975    3rd Qu.: 0.77500
 Max.   : 2.63000    Max.   : 2.7000    Max.   : 2.73000
    mathmtcs              geometry             analyrea
 Min.   :-3.7100    Min.   :-3.3200    Min.   :-2.8300
 1st Qu.:-0.4925    1st Qu.:-0.5600    1st Qu.:-0.4825
 Median : 0.1000    Median : 0.0900    Median : 0.2000
 Mean   : 0.1055    Mean   : 0.1125    Mean   : 0.1750
 3rd Qu.: 0.9200    3rd Qu.: 0.7675    3rd Qu.: 0.8375
 Max.   : 3.0600    Max.   : 3.8600    Max.   : 3.5000
```

```
apply(cognition, 2, var)
```

```
    vocab   reading  sentcomp  mathmtcs  geometry  analyrea
0.9966514 0.9811568 0.9834142 1.1117325 1.0686631 1.1170926
```

Looking at the descriptive statistics, it appears that the variables may have been standardized (indicated by the means of the variables being approximately zero and the variances being approximately one). More likely, because this is a simulated data set, the variables were randomly generated from a standard normal distribution. The means and medians are all close to one another, suggesting the distributions may be reasonably symmetric. We see that the mathmtcs and geometry variables have the largest ranges and that there are no missing data because there are no NAs printed from summary(cognition). We can glean much more information about the distributions of these indicators by plotting the variables.

To start, we look at histograms of the marginal distributions of our measures of cognition using the **lattice** package (Sarkar, 2008). We first load the **lattice** package. Then we use the reshape function to transform the data from wide format into long format to make plotting the marginal distributions easier. In the wide format, each row represents a single person and each column represents a different variable related to the person (e.g., age, gender, test scores). Unlike the wide format, in the long format each person has multiple rows with each row corresponding to a score on a given variable in the `cognition` data set.

In the following example, we transform the wide format `cognition` data set into a long format data set and save this new data set as `cognition_1`. Using the `head` function, we print the first six rows of the long format data set after sorting the data by participant identifier "id."

```
library("lattice")
cognition_1 <- reshape(data = cognition,
                       varying = 1:6,
                       v.names = "score",
                       timevar = "indicator",
                       times = names(cognition),
                       direction = "long")
head(cognition_1[order(cognition_1$id),])
```

```
          indicator score id
1.vocab        vocab  1.67  1
1.reading    reading  1.67  1
1.sentcomp  sentcomp  1.46  1
1.mathmtcs  mathmtcs  0.90  1
1.geometry  geometry  0.49  1
1.analyrea  analyrea  1.65  1
```

Next, we use the `histogram` function specifying the variable that contains the scores for each indicator (i.e., score). In the `histogram` function, we use the | to create a trellis (or facet) plot by indicator. This will create a histogram for each indicator and arrange them in a grid (see Chapters 1 and 2 for more information about these plots).

```
histogram(~ score | indicator, data = cognition_1)
```

In Figure 4.1, we see that the variables are unimodal, symmetric, and roughly normal. In these plots, we see the presence of some extreme observations (particularly for the mathematical indicators). Next, we examine the bivariate relationships between our indicators using a scatter plot matrix. In a scatter plot matrix, we typically look for four things:

1. the magnitude of the relationships between the variables,

2. the direction of the relationships between the variables,

3. whether there are extreme or potentially influential points, and

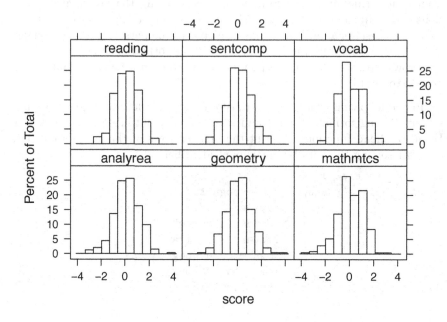

FIGURE 4.1
Histograms of indicators of cognition.

4. the presence of non-linear relationships.

To create a scatter plot matrix for the `cognition` data set, we use the `pairs` function as shown below:

```
pairs(cognition)
```

In the scatter plot matrix shown in Figure 4.2, we see that all of the variables are positively correlated with one another (based on the increasing, positive trend). It appears that the verbal measures (vocab, reading, and sentcomp) and the math measures (mathmtcs, geometry, and analyrea) are more strongly correlated with one another. This suggests that there might be two factors; however, the variables are all at least moderately related to one another, suggesting that one general factor might be sufficient. In most of the plots in Figure 4.2, we see the presence of extreme values or outliers, but we do not see any evidence of nonlinear relationships between the variables.

Finally, we look at a plot to detect multivariate influential cases that was described in Bollen (1989). This plot tells us roughly how far a particular observation is from the centroid of the data (more information about visualizing data can be found in Chapter 9). This plot can be created using the

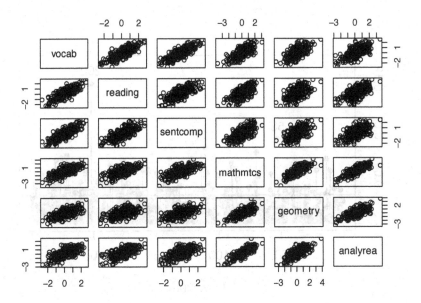

FIGURE 4.2
Scatterplot matrix of the cognition measures.

`bollen_plot` function in the **hemp** package. The `bollen_plot` function accepts two arguments, a data frame (required) and `crit.value`, which is an optional argument to label the row numbers that correspond to values above a certain threshold.

```
bollen_plot(cognition, crit.value = .06)
```

In Figure 4.3, the black horizontal line corresponds to the expected size, 0.024 (the number of manifest variables divided by the number of observations), so large values relative to the expected size could be considered influential (see Bollen (1989) for more details). We see that there are a handful of observations that are quite far from the others. In addition, the values greater than 0.06 (based on the `crit.value` argument that we specified) have their row numbers labeled. Row numbers 202, 53, and 111 are observations that may exert substantial influence on our findings. Below, we look at these observations, as they could represent data that were entered incorrectly, or values that represent a part of our population that we may have undersampled.

```
cognition[c(202, 53, 111),]
```

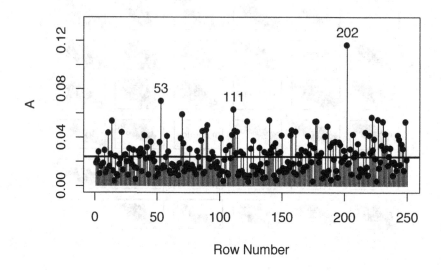

FIGURE 4.3
Bollen plot of cognition data.

	vocab	reading	sentcomp	mathmtcs	geometry	analyrea
202	2.63	2.23	2.55	1.38	3.86	3.50
53	-0.38	0.99	-0.50	1.79	-0.19	2.13
111	-2.01	-2.47	-2.47	-3.71	-2.59	-2.67

The code above prints out the row numbers (i.e., participants) 202, 53, and 111 from the `cognition` data set. Then, below, we print the minimum and maximum values for each indicator.

`apply(cognition, 2, min)`

vocab	reading	sentcomp	mathmtcs	geometry	analyrea
-2.62	-2.47	-2.47	-3.71	-3.32	-2.83

`apply(cognition, 2, max)`

vocab	reading	sentcomp	mathmtcs	geometry	analyrea
2.63	2.70	2.73	3.06	3.86	3.50

We see that observation 202 has the maximum observed value in `vocab`, `geometry`, and `analyrea` but has a slightly lower `mathmtcs` score than might be expected given their other scores. Similarly, observation 53 has a high analyrea score but a low geometry score. Finally, 111 is, generally, a poor performer. Given that the data were fabricated, there is no reason to believe that the data were keyed in incorrectly and while a score of 1.38 for an individual obtaining the maximum geometry and analyrea scores is unlikely, it is not impossible.

Only after looking at these statistics, or similar plots, we recommend looking at the Pearson correlations between the indicators. Looking at plots first would allow us to assess the appropriateness of the Pearson correlation. Below we calculate the Pearson correlations using the cor function, save the results as correlations, and finally, round them off to the nearest thousandth using the round function.

```
correlations <- cor(cognition)
round(correlations, 3)
```

```
         vocab reading sentcomp mathmtcs geometry analyrea
vocab    1.000  0.803   0.813    0.708    0.633    0.673
reading  0.803  1.000   0.725    0.660    0.526    0.636
sentcomp 0.813  0.725   1.000    0.618    0.575    0.618
mathmtcs 0.708  0.660   0.618    1.000    0.774    0.817
geometry 0.633  0.526   0.575    0.774    1.000    0.715
analyrea 0.673  0.636   0.618    0.817    0.715    1.000
```

The correlations support what we have seen in Figure 4.2. The variables in the cognition data set are all moderately to strongly correlated with one another, suggesting that either a one-factor or two-factor structure might be sufficient. Within a category of indicators (i.e., either verbal or mathematics), we see that the indicators are more highly correlated with one another than across the categories.

As a sensitivity analysis, we also recommend calculating the differences in the correlation matrix using all the observations and the correlation matrix dropping just observations 202, 53, and 111.

```
cor_diff <- correlations - cor(cognition[-c(202,53,101),])
round(cor_diff, 3)
```

```
          vocab reading sentcomp mathmtcs geometry analyrea
vocab     0.000  0.001   0.005   -0.008    0.008    0.002
reading   0.001  0.000   0.004    0.000    0.002    0.010
sentcomp  0.005  0.004   0.000   -0.010    0.004    0.002
mathmtcs -0.008  0.000  -0.010    0.000   -0.012   -0.002
geometry  0.008  0.002   0.004   -0.012    0.000    0.002
analyrea  0.002  0.010   0.002   -0.002    0.002    0.000
```

We find that the difference between the correlation matrices is extremely small, with the largest difference occurring between the geometry and `mathmtcs` variables (-.012). This finding indicates that the influential observations are impacting the correlations minimally.

After all of these preliminary analyses, we are now ready to fit an EFA model to the `cognition` data set. There are several criteria that we need to consider when determining how many factors to extract from a factor analysis. The simplest approach involves examining the eigenvalues from an eigen decomposition of a correlation matrix of the variables. We run an eigen decomposition on `correlations`, which we save as `eigen_decomp`, and then we print the eigenvalues rounded to the nearest thousandth.

```
eigen_decomp <- eigen(correlations)
round(eigen_decomp$values, 3)
```

```
[1] 4.436 0.676 0.322 0.245 0.168 0.152
```

From the output, we see that there is one large eigenvalue and if we used Kaiser's rule of extracting the number of factors to be equal to the number of eigenvalues greater than one (Kaiser, 1960), then we would consider just one factor. If we fit PCA on this dataset, we can see that the first component alone would explain about 74% of the observed variability.

Another common method involves constructing a scree plot and looking for the location of the elbow in the plot, and extracting the number of factors to be equal to the eigenvalues occurring right before the elbow. The following example demonstrates how to create a scree plot using the `lattice_scree` function in the **hemp** package.

```
lattice_scree(cognition)
```

Again, we see evidence of one major factor and potentially a second factor as the eigenvalues begin to level off after two factors. Note that this is a scree plot using the eigenvalues of the reduced correlation matrix, and eigenvalues greater than 0 would indicate the number of factors we might consider extracting. A more sophisticated approach would involve using parallel analysis (Horn, 1965). Parallel analysis can be done using the `fa.parallel` function in the **psych** package as demonstrated below. We need to specify the name of the data set with the variables of interest (`cognition`) and the estimation method (`fm = "ml"` for maximum likelihood) .

```
library("psych")
fa.parallel(cognition, fm = "ml")
```

We are not going to review the results from `fa.parallel`. Instead, we will use an additional function in the **hemp** package called `lattice_pa` to perform and plot a parallel analysis. The `lattice_pa` function uses a modified version of the `fa.parallel` function, which is appropriate to use when we want to use maximum likelihood as the discrepancy function and the data

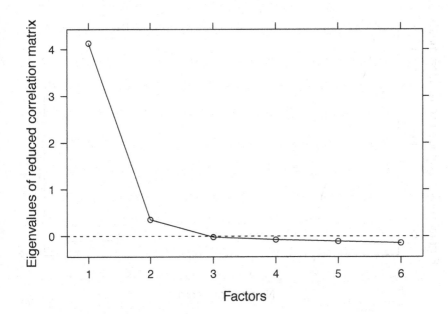

FIGURE 4.4
Scree plot of the cognition dataset.

are ratio or interval scale. Otherwise, we recommend using the `fa.parallel` function if the variables are not ratio or interval scale; a different discrepancy function (e.g., minimum residual, principal axis, and weighted least squares) is necessary; or finer control over the parallel analysis is desired.

```
lattice_pa(cognition)
```

The solid line corresponds to eigenvalues of the reduced correlation matrix (i.e., by replacing the ones in the diagonal of the correlation matrix with the estimated commonalities), and the dotted line corresponds to the eigenvalues of a reduced correlation matrix based on simulated random data. To use parallel analysis, we would extract the number of factors that occur above the dotted line. Based on the parallel analysis, we would extract two factors. We will consider both one factor and two factor solutions below.

We begin by fitting a one factor solution to the `cognition` data set. To use the `factanal` function, we need to specify the name of the data set (or a covariance matrix) and the number of factors to extract (i.e., `factors = 1`). We save the results as `one_factor` and then print the output.

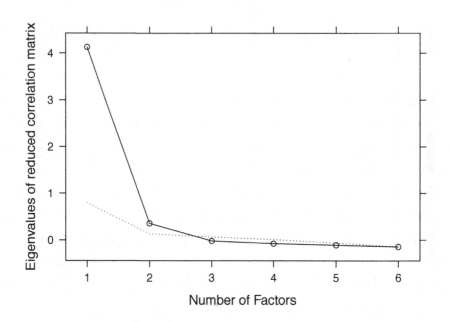

FIGURE 4.5
Parallel analysis of the cognition dataset.

```
one_factor <- factanal(cognition, factors = 1)
one_factor
```

```
Call:
factanal(x = cognition, factors = 1)

Uniquenesses:
  vocab  reading sentcomp mathmtcs geometry analyrea
  0.213    0.324    0.331    0.270    0.412    0.318

Loadings:
         Factor1
vocab    0.887
reading  0.822
sentcomp 0.818
mathmtcs 0.854
geometry 0.767
analyrea 0.826
```

```
                Factor1
SS loadings      4.132
Proportion Var   0.689


Test of the hypothesis that 1 factor is sufficient.
The chi square statistic is 171.91 on 9 degrees of freedom.
The p-value is 2.46e-32
```

First, the call is printed (i.e., what we told `factanal` to run); then the uniquenesses, which are the proportion of the variability that are unexplained by the extracted factor; then the factor loadings, which in this case represent the correlations of the indicators with the factor, and when squared represent the proportion of variability in the indicators explained by the factor (i.e., the commonalities); the sum of the factor loadings squared (SS loadings); the proportion of variability explained by the factor (which is also equal to the SS loadings divided by the number of indicators); and finally, a chi-square statistic is printed where the null hypothesis is that the one-factor solution is sufficient.

In the output, we see that all of the indicators loaded highly on the extracted factor and that this one factor is able to explain approximately 69% of the variability in the `cognition` data. Note that we reject the chi-square statistic, which suggests that a one-factor solution is not sufficient. Given this, we next consider a two-factor solution. We update the code by specifying two factors with `factors = 2`. The results are saved as `two_factor`.

```
two_factor <- factanal(cognition, factors = 2)
two_factor
```

```
Call:
factanal(x = cognition, factors = 2)

Uniquenesses:
   vocab  reading sentcomp mathmtcs geometry analyrea
   0.104    0.281    0.261    0.109    0.327    0.246

Loadings:
         Factor1 Factor2
vocab     0.853   0.410
reading   0.750   0.395
sentcomp  0.785   0.350
mathmtcs  0.424   0.843
geometry  0.384   0.725
analyrea  0.437   0.751


         Factor1 Factor2
```

```
SS loadings       2.426   2.246
Proportion Var    0.404   0.374
Cumulative Var    0.404   0.779


Test of the hypothesis that 2 factors are sufficient.
The chi square statistic is 11.22 on 4 degrees of freedom.
The p-value is 0.0242
```

In the two-factor solution, we see that `vocab`, `reading`, and `sentcomp` load highly on the first factor while `mathmtcs`, `geometry`, and `analyrea` load highly on the second factor. The potential names for the first and second factors could be verbal reasoning and mathematical reasoning, respectively. The two-factor structure explains 78% of the variability in the `cognition` data set. With the two-factor structure, we still reject the null hypothesis and conclude that the two-factor structure is not sufficient based on the chi-square test. This test is often rejected and not necessarily a robust test of the adequacy of the proposed factor structure. We note that the indicators, while loading more highly on one factor than another, cross load moderately, resulting in a non-simple structure. The default rotation for `factanal` is varimax, which is an orthogonal rotation that forces the factors to be uncorrelated. We wish to achieve a simple structure, where each indicator is associated with only a single factor, and therefore, it may be beneficial to consider an oblique rotation.

To perform an oblique rotation in R, we recommend using `factanal` with the default rotation and then rotating the solution because of a long standing bug.[1] A variety of orthogonal and oblique rotations are available in the **GPArotation** package (Bernaards & Jennrich, 2005). To perform a rotation, we specify the name of the rotation that we want to use and pass it the factor loadings from the unrotated solution. In the example below, we install and load the **GPArotation** package, and then perform an oblimin rotation passing the loadings from the `two_factor` object that we created earlier.

```
install.packages("GPArotation")
library("GPArotation")
oblimin(loadings(two_factor))
```

```
Oblique rotation method Oblimin Quartimin converged.
Loadings:
          Factor1  Factor2
vocab     0.94412  0.00325
reading   0.80339  0.05482
sentcomp  0.88899 -0.03725
mathmtcs -0.03159  0.96884
geometry  0.00144  0.81923
```

---

[1] https://bugs.r-project.org/bugzilla3/show_bug.cgi?id=15848

```
analyrea  0.05990  0.81970

Rotating matrix:
       [,1]    [,2]
[1,]  1.483 -0.723
[2,] -0.783  1.513

Phi:
       [,1]   [,2]
[1,]  1.000  0.803
[2,]  0.803  1.000
```

We see that oblimin rotation allowed us to obtain a simple structure (i.e., all the indicators load highly on one factor and close to zero on the other) and that the correlation between the two factors was 0.803 (Phi is the correlation matrix between the factors). In this case, we find that simply allowing the factors to be correlated removed the cross loading.

Factor scores can be extracted by specifying the scores argument. Thompson's scores can be extracted using scores = "regression" and Bartlett's weighted least-squares scores can be extracted using scores = "Bartlett" (see Bartholomew, Deary, and Lawn (2009) for a review of these factor scoring methods). Caution should always be used with factor scores because of indeterminacy (Mulaik, 2009).

4.3.2 EFA Using the psych Package

In the event that our data violate multivariate normality, we may want to investigate a different statistical discrepancy function for EFA. The **psych** package offers several options including using OLS to find the minimum residual solution (the default statistical discrepancy function). In the following example, we fit an EFA model to the same cognition data set using the fa function in the **psych** package. We specify the number of factors as 2 (nfactors = 2) and the rotation method as oblique rotation (rotate = "oblimin").

```
library("psych")
fa(r = cognition, nfactors = 2, rotate = "oblimin")
```

```
Factor Analysis using method =  minres
Call: fa(r = cognition, nfactors = 2, rotate = "oblimin")
Standardized loadings (pattern matrix) based upon correlation
matrix
          MR1   MR2   h2   u2 com
vocab    0.95  0.00 0.90 0.10   1
reading  0.83  0.02 0.72 0.28   1
```

```
sentcomp  0.87 -0.02 0.74 0.26   1
mathmtcs -0.03  0.96 0.89 0.11   1
geometry -0.02  0.84 0.68 0.32   1
analyrea  0.07  0.81 0.76 0.24   1

                        MR1  MR2
SS loadings            2.36 2.31
Proportion Var         0.39 0.38
Cumulative Var         0.39 0.78
Proportion Explained   0.51 0.49
Cumulative Proportion  0.51 1.00

 With factor correlations of
    MR1 MR2
MR1 1.0 0.8
MR2 0.8 1.0

Mean item complexity = 1
Test of the hypothesis that 2 factors are sufficient.

The degrees of freedom for the null model are  15  and the
objective function was  5.11 with Chi Square of  1257.24

The degrees of freedom for the model are 4  and the
objective function was  0.05

The root mean square of the residuals (RMSR) is  0.01
The df corrected root mean square of the residuals is  0.03

The harmonic number of observations is  250 with the
empirical chi square  1.42  with prob <  0.84

The total number of observations was  250  with
Likelihood Chi Square =  11.64  with prob <  0.02

Tucker Lewis Index of factoring reliability =  0.977
RMSEA index =  0.089
and the 90 % confidence intervals are  0.031 0.148

BIC =  -10.44
Fit based upon off diagonal values = 1
Measures of factor score adequacy
                                           MR1  MR2
Correlation of scores with factors         0.97 0.97
```

```
Multiple R square of scores with factors       0.94 0.94
Minimum correlation of possible factor scores  0.88 0.87
```

The output from the **fa** function is a bit more verbose compared to the output from **factanal** because it includes information about fit indices. The first table contains the factor loadings (columns 1 and 2), the commonalities, the uniquenesses, and the sum of these two columns. The second table prints the sum of squared loadings for each factor, proportion of variability explained for each factor, cumulative variance explained for each factor, proportion of the total explained variability accounted for by each factor, and the cumulative proportion of explained variability accounted for by each factor (this must sum to 1). Then a table of factor correlations and two chi-square tests, the empirical chi-square statistic and the maximum likelihood chi-square statistic, are printed. Finally, we see fit measures including RMSR (root mean square residuals), Tucker Lewis Index (TLI), and RMSEA (root mean square error of approximation). Overall, the fit measures suggest that the two-factor model fits the **cognition** data set well (based on commonly used criteria, such as Hu and Bentler (1999)). Based on this analysis, we found evidence supporting a two-factor structure for the indicators in the **cognition** data set.

4.3.3 EFA with Categorical Data

The examples so far have focused on continuous variables on interval or ratio scales of measurement. In this section, we show how to perform EFA using categorical data with the **psych** package. We will use a subset of the items from the **SAPA** data set, which was described in Chapter 2. We save this new data set as **SAPA_subset**.

```
SAPA_subset <- subset(SAPA, select = c(letter.7:letter.58,
                                       rotate.3:rotate.8))
```

The variables in the **SAPA_subset** data set are dichotomously scored (i.e., 0 and 1 as the values for the variables). Therefore, when we run a parallel analysis or factor analysis, we need to specify polychoric correlations instead of the Pearson correlation (the default option for the **fa.parallel** and **fa** functions) because the Pearson correlations would be attenuated because of the discretization. A parallel analysis can be performed by running the following:

```
fa.parallel(SAPA_subset, cor = "poly")
```

The results of the parallel analysis suggest that 2 factors or 2 components could be extracted from the data. We also perform EFA on the same data set using the **fa** function in **psych** by specifying the cor = "poly" argument.

```
EFA_SAPA <- fa(SAPA_subset, 2, rotate = "oblimin", cor = "poly")
EFA_SAPA
```

```
              MR1    MR2    h2    u2  com
letter.7   -0.02   0.79  0.60  0.40  1.0
letter.33   0.01   0.70  0.50  0.50  1.0
letter.34  -0.02   0.80  0.63  0.37  1.0
letter.58   0.21   0.54  0.46  0.54  1.3
rotate.3    0.86  -0.02  0.72  0.28  1.0
rotate.4    0.82   0.10  0.77  0.23  1.0
rotate.6    0.77   0.05  0.64  0.36  1.0
rotate.8    0.86  -0.09  0.65  0.35  1.0

                          MR1   MR2
SS loadings              2.84  2.13
Proportion Var           0.36  0.27
Cumulative Var           0.36  0.62
Proportion Explained     0.57  0.43
Cumulative Proportion    0.57  1.00

With factor correlations of
     MR1   MR2
MR1 1.00  0.56
MR2 0.56  1.00

The root mean square of the residuals (RMSR) is  0.02
Tucker Lewis Index of factoring reliability =  0.972
RMSEA index =  0.065
```

Note that we only printed a portion of the output returned from the fa function. From the output, we can see that we have achieved a simple structure using the oblimin rotation with reasonably strong evidence of a two-factor structure (two factors account for approximately 62% of the variability, RMSR is .02, TLI is .972, and RMSEA is .065). We could also use the factanal function directly by passing a polychoric correlation matrix as the input instead of the SAPA_subset data set. To accomplish this, we first calculate the polychoric correlations using the polychoric function and then pass this to the factanal function (covmat = SAPA_cor$rho).

```
SAPA_cor <- polychoric(SAPA_subset)
factanal(covmat = SAPA_cor$rho, factors = 2, n.obs = nrow(SAPA))
```

4.4   Confirmatory Factor Analysis (CFA)

After an instrument has been developed and validated, we have a sense of the dimensionality of the instrument and which indicators should load onto which factor(s). In this setting, it is appropriate to consider CFA for examining the factor structure, not EFA. Unlike with EFA, in CFA we place certain restrictions on the model-implied matrices in Equations 4.1 and 4.2. In CFA, we do not concern ourselves with rotation and indeterminacy because we fix our scale by setting a single factor loading for each factor or the variance of the factors to 1.

4.4.1   CFA of the WISC-R Data

The WISC-R is a revised version of the WISC (a downward extension of the Weschler-Belleview Test for Children) published in 1975. The data come from an administration of the WISC-R to 175 children. Details about the data can be found in Tabachnick, Fidell, and Osterlind (2001). This data set is included in the **hemp** package as `wiscsem`. Variables in `wiscsem` include client (an id variable), agemat (a categorical age variable), scores on ten core subtests, and scores on one optional subtest.

For brevity, we will skip all the initial exploratory descriptive and graphical analyses that we previously performed while doing the EFA. However, this should be done with CFA as well as any type of psychometric analysis. To perform the actual CFA in R, as well as path analysis and structural equation modeling (SEM), we will use the **lavaan** package Rosseel (2012), which stands for **l**atent **va**riable **an**alysis. Below we install and activate the **lavaan** package.

```
install.packages("lavaan")
library("lavaan")
```

The **lavaan** package is capable of regression, CFA, SEM, growth curve modeling, mediation, moderation, and much more. Using **lavaan** involves the following three steps:

1. defining the model using **lavaan**'s special syntax (similar to Mplus),

2. fitting the model, and

3. extracting information from the fitted model.

The **lavaan** package uses its own special model syntax. For conducting a CFA, the following syntax is typically used:

=~ : This is how we define our latent variables and the indicators associated with them.

~~ : This is how we define a covariance or variance.

~ : This is how we define a path for path analysis, regression, or SEM.

+ : This is how we string together variables (identical to how we would fit multiple regression with lm).

The author of **lavaan** has a great website, with well-documented tutorials, and we refer interested readers to this website for more details about **lavaan**.[2]

Because we know which subtests were supposed to measure which component of intelligence (i.e., either verbal (verb) or performance (perf) IQ), we define our model to reflect this structure. The first thing we need to do is define the model using **lavaan**'s syntax.

```
iq_mod <- '
  verb =~ info + comp + arith + simil + digit + vocab
  perf =~ pictcomp + parang + block + object + coding
'
```

By default, **lavaan** fixes the first indicator on each latent variable, info and pictcomp in this case, to "1" to set the scale for the factors. We could have set the scales of verb and perf to 1 to obtain a standardized solution. However, it is easier to request the standardized solution from the summary function (shown below). Therefore, we recommend sticking with the default options in this simple case. Next, we fit the model (iq_mod) using the **cfa** function in **lavaan**. We could use the lavaan function if we want fine control over the model-fitting procedure (which can be quite useful for running simulations and fitting more complex, structural equation models). Because we are fitting a relatively simple CFA model, we will use the **cfa** function to estimate the model and save it as an object called iq_fit.

```
iq_fit <- cfa(iq_mod, data = wiscsem)
```

The first thing we recommend doing is inspecting the model to verify that **lavaan** actually fit the model that we intended to fit. For this step, we use the CFA results iq_fit with the **inspect** function.

```
inspect(iq_fit)
```

```
$lambda
        verb perf
info       0    0
comp       1    0
arith      2    0
```

---

[2]http://lavaan.ugent.be/

```
simil        3    0
digit        4    0
vocab        5    0
pictcomp     0    0
parang       0    6
block        0    7
object       0    8
coding       0    9
```

$theta

	info	comp	arith	simil	digit	vocab	pctcmp	parang
info	10							
comp	0	11						
arith	0	0	12					
simil	0	0	0	13				
digit	0	0	0	0	14			
vocab	0	0	0	0	0	15		
pictcomp	0	0	0	0	0	0	16	
parang	0	0	0	0	0	0	0	17
block	0	0	0	0	0	0	0	0
object	0	0	0	0	0	0	0	0
coding	0	0	0	0	0	0	0	0

	block	object	coding
info			
comp			
arith			
simil			
digit			
vocab			
pictcomp			
parang			
block	18		
object	0	19	
coding	0	0	20

$psi

	verb	perf
verb	21	
perf	23	22

The $lambda matrix corresponds to the factor loadings that are being estimated for the indicators. A non-zero value indicates that the loading is being estimated for that factor. Note that info and pictcomp are both zero. This is because we fixed their loadings to 1. The $theta matrix is the covariance matrix of our indicators, which refers to the residual variances for our indicators (i.e., the unique or specific variance). All the indicators have residuals estimated and none of the indicators are correlated (i.e., all off-diagonal elements are zero). The $psi matrix is the covariance matrix of the factors. All elements of this matrix (i.e., variances and covariances) are being estimated. In other words, we assume that verbal and performance intelligence are correlated.

Based on the results from the `inspect` function, we can see that the CFA model is indeed what we intended to estimate. Now we can use the `summary` function to extract information from our fitted model.

```
summary(iq_fit)
```

```
lavaan (0.5-23.1097) converged normally after  51 iterations

  Number of observations                          175

  Estimator                                        ML
  Minimum Function Test Statistic              70.640
  Degrees of freedom                               43
  P-value (Chi-square)                          0.005

Parameter Estimates:

  Information                                 Expected
  Standard Errors                             Standard

Latent Variables:
                   Estimate  Std.Err  z-value  P(>|z|)
  verb =~
    info              1.000
    comp              0.926    0.108    8.609    0.000
    arith             0.589    0.084    7.013    0.000
    simil             1.012    0.115    8.764    0.000
    digit             0.477    0.099    4.805    0.000
    vocab             1.020    0.107    9.548    0.000
  perf =~
    pictcomp          1.000
    parang            0.719    0.156    4.614    0.000
    block             1.060    0.187    5.675    0.000
    object            0.921    0.177    5.215    0.000
    coding            0.119    0.147    0.810    0.418

Covariances:
                   Estimate  Std.Err  z-value  P(>|z|)
  verb ~~
    perf              2.263    0.515    4.397    0.000
```

Variances:

	Estimate	Std.Err	z-value	P(>|z|)
.info	3.566	0.507	7.034	0.000
.comp	4.572	0.585	7.815	0.000
.arith	3.602	0.420	8.571	0.000
.simil	5.096	0.662	7.702	0.000
.digit	6.162	0.680	9.056	0.000
.vocab	3.487	0.506	6.886	0.000
.pictcomp	5.526	0.757	7.296	0.000
.parang	5.463	0.658	8.298	0.000
.block	3.894	0.640	6.083	0.000
.object	5.467	0.719	7.600	0.000
.coding	8.159	0.874	9.335	0.000
verb	4.867	0.883	5.514	0.000
perf	3.035	0.844	3.593	0.000

The output is intended to be similar to Mplus (Muthén & Muthén, 2015). However, the default output is a bit more terse. The Minimum Function Test Statistic is the chi-square statistic for the current model and is a measure of deviance between the model-implied covariance matrix and the observed covariance matrix. Next is the number of degrees of freedom, followed by the p-value of this test. The degrees of freedom can be calculated as the number of unique elements in the covariance matrix minus the number of free parameters using the unique_elements and free_params functions in the **hemp** package.

```
unique_elements(iq_fit) - free_params(iq_fit)
```

```
[1] 43
```

The summary(iq_fit) argument returns three tables. The first table corresponds to the estimated factor loadings, with standard errors, z-values, and p-values. The second table shows the covariances of the factors, and the last table contains the variances of both the indicators and the factors (the factors are listed last). The most noteworthy finding in these tables is that coding appears to be unrelated to performance IQ. It should be noted that the **cfa** function does not provide intercepts for the indicators by default (unlike Mplus). However, intercepts can be included in the model using the meanstructure = TRUE argument. These intercept terms correspond to the raw means of the indicators.

Although the summary function, by default, returns the unstandardized solution for the estimated CFA model, it is possible to print the standardized solution as well as model fit measures by including standardized = TRUE and fit.measures = TRUE in the summary function.

```
summary(iq_fit, standardized = TRUE, fit.measures = TRUE)
```

The `standardized` = TRUE argument adds two extra columns to the co-efficient tables. The Std.lv column is the standardized output based on only standardizing the latent variable and the Std.all column is the standardized output based on standardizing the latent variable and the indicator. The former is useful when the indicators are discrete, while the latter is useful if the indicators are continuous.

The `fit.measures` = TRUE argument prints numerous fit indices (e.g., Bayesian information criteria, Akaike's information criteria, the comparative fit index (CFI), TLI, RMSEA, and SRMR). It also provides an additional chi-square test of the most parsimonious, plausible model that estimates only the variances of the indicators. A variety of additional fit indices can be extracted from the fitted model. To print all the available indices, we can use the `fitMeasures` function.

```
fitmeasures(iq_fit)
```

We can use the `inspect` function to print the R^2 (the proportion of variability in our indicators that are explained by the latent variable(s)) as a separate output.

```
inspect(iq_fit, "rsquare")
```

```
    info     comp    arith    simil    digit    vocab
   0.577    0.477    0.319    0.494    0.152    0.592
 pictcomp   parang    block   object   coding
   0.354    0.223    0.467    0.320    0.005
```

The same information could also be obtained from the default summary output (i.e., 1 - unique variance). We can also pass an additional argument to the `summary` function to request the R^2.

```
summary(iq_fit, rsquare = TRUE)
```

Returning to our CFA model, we drop the variable "coding" from the model and fit the model again. We save the results from this new model as `iq_fit_nocode`.

```
iq_mod_nocode <- '
  verb =~ info + comp + arith + simil + digit + vocab
  perf =~ pictcomp + parang + block + object
'

iq_fit_nocode <- cfa(iq_mod_nocode, wiscsem)
```

Next, we can compare the model with coding (`code`) to the one without coding (`nocode`) and combine the fit indices using the `rbind` function.

```
nocode <- fitmeasures(iq_fit_nocode,
                fit.measures = c("rmsea", "tli", "cfi"))
code <- fitmeasures(iq_fit,
                fit.measures = c("rmsea", "tli", "cfi"))
rbind(nocode, code)
```

```
           rmsea        tli        cfi
nocode 0.05983914 0.9384618 0.9535045
code   0.06060555 0.9238418 0.9404581
```

These models are not nested as they have different covariance matrices and we cannot compare these models with a chi-square test of difference or information criterion. However, we can compare their absolute fit indices. Models are only nested when they have the same covariance matrix but with fewer free parameters. In that situation, we can simply use a chi-square test of difference.

Based on this output, we could justify dropping coding but dropping coding does not seem to improve model fit drastically. If we want to improve model fit, we could consider examining modification indices that can be calculated using the `modindices` function. Our present model is nested within the models where all these additional paths are free. Below we calculate the modification indices, sort them in decreasing order, and print only those that would significantly improve model fit based on the chi-square test (i.e., the changes in χ^2 square would be greater than 3.84 with df = 1).

```
modind_iq <- modindices(iq_fit_nocode, sort. = TRUE)
modind_iq[modind_iq$mi > 3.84,]
```

	lhs	op	rhs	mi	epc	sepc.lv	sepc.all	sepc.nox
29	perf	=~	comp	9.853	0.534	0.939	0.318	0.318
57	arith	~~	object	6.246	-0.936	-0.936	-0.144	-0.144
34	info	~~	comp	5.247	-0.987	-0.987	-0.115	-0.115
47	comp	~~	pictcomp	4.629	0.961	0.961	0.111	0.111
38	info	~~	vocab	4.429	0.915	0.915	0.108	0.108
28	perf	=~	info	4.372	-0.335	-0.589	-0.203	-0.203
35	info	~~	arith	4.156	0.699	0.699	0.105	0.105
70	vocab	~~	parang	4.072	-0.791	-0.791	-0.102	-0.102

In the output, the first three columns correspond to the paths to be freed. The op column corresponds to the **lavaan** operator between the variable on the left-hand side (lhs) and the variable on the right-hand side (rhs). Earlier we saw that =~ means that the variable on the left is manifested by the variable on the right. The first line indicates that we should allow comp to be a manifestation of the performance IQ variable perf. The second line includes the ~~ operator, suggesting that we should allow the variable on the left and the variable on the right to covary. In this case, we should allow arith to covary with object.

The remaining columns are the expected change in the chi-square statistic; the expected parameter change (i.e., the estimated parameter of that free path); and then standardized versions of the following: standardizing just the latent variable, standardizing all the variables, and standardizing all but the exogenous variables in the model. The modification indices (i.e., the mi column) show that the biggest change in model fit would be found by allowing comp to load onto perf. The change in chi-square would be 9.823.

Using the suggestions based on modification indices, we update the CFA model by adding comp under the factor perf and save the new model as iq_fit_mi.

```
iq_mod_mi <- '
  verb =~ info + comp + arith + simil + digit + vocab
  perf =~ pictcomp + parang + block + object + comp
'
iq_fit_mi <- cfa(iq_mod_mi, wiscsem)
fitMeasures(iq_fit_mi, c("rmsea", "tli", "cfi", "srmr"))
```

```
rmsea   tli   cfi  srmr
0.046 0.963 0.973 0.049
```

Now we see that the fit measures are very good in the updated model (i.e., iq_fit_mi), and if we examined the summary output, we would see that we fail to reject the null hypothesis for the chi-square test. However, we should be ready to justify why we would allow comp to be a manifestation of perf. That is a validity question and has ramifications for generalizability as well (i.e., are we overfitting our model?). Because the models are now nested, we can use the anova function to perform a chi-square test of difference.

```
anova(iq_fit_nocode, iq_fit_mi)
```

```
Chi Square Difference Test

              Df    AIC    BIC  Chisq Chisq diff Df diff
iq_fit_mi     33 8153.4 8223.0 45.276
iq_fit_nocode 34 8161.4 8227.9 55.305     10.029      1
              Pr(>Chisq)
iq_fit_mi
iq_fit_nocode   0.001541 **
---
Signif. codes:
0 '***' 0.001 '**' 0.01 '*' 0.05 '.' 0.1 ' ' 1
```

We see that allowing comprehension to load on performance IQ results in a statistically significant improvement in model fit. To extract the factor scores from our best-fitting model, `iq_fit_mi`, we use the `predict` function, save the results as `factor_scores`, and print the first six rows using the `head` function.

```
factor_scores <- predict(iq_fit_mi)
head(factor_scores)
```

```
            verb       perf
[1,] -0.2227813 -1.2256055
[2,] -1.1473545 -2.0686206
[3,]  3.8269042  2.4161738
[4,] -0.9696807 -0.8159658
[5,] -0.9728589 -2.5860298
[6,]  0.8370746 -1.7503289
```

Finally, if we want to extract the model-implied covariance and the residual covariance matrices from a CFA model estimated with the `cfa` function, we can use the `fitted` and the `residual` functions.

```
fitted(iq_fit_mi)
```

```
$cov
          info   comp  arith simil digit vocab pctcmp
info      8.433
comp      4.429  8.743
arith     2.948  2.564  5.291
simil     4.976  4.329  2.881 10.078
digit     2.424  2.108  1.403  2.369 7.271
vocab     5.129  4.461  2.970  5.013 2.442 8.552
pictcomp  2.172  3.035  1.258  2.123 1.034 2.188  8.560
parang    1.426  1.993  0.826  1.394 0.679 1.437  2.144
block     2.177  3.042  1.260  2.128 1.036 2.193  3.272
object    1.961  2.741  1.136  1.917 0.934 1.976  2.949
          parang block  object
info
comp
arith
simil
digit
vocab
pictcomp
parang    7.033
block     2.148  7.301
object    1.936  2.955  8.042

$mean
```

```
       info     comp    arith    simil    digit    vocab
          0        0        0        0        0        0
    pictcomp   parang    block   object
          0        0        0        0
```

```
residuals(iq_fit_mi)
```

```
$type
[1] "raw"
```

```
$cov
            info   comp   arith  simil  digit  vocab  pctcmp
info        0.000
comp       -0.417  0.000
arith       0.355  0.104  0.000
simil      -0.246  0.460 -0.184  0.000
digit       0.281 -0.228  0.265 -0.148  0.000
vocab       0.179  0.133 -0.364 -0.020 -0.121  0.000
pictcomp   -0.219  0.485 -0.212  1.307 -0.440  0.254  0.000
parang      0.126 -0.530  0.557  1.116  0.381 -0.412 -0.214
block      -0.379 -0.093  0.431  0.115 -0.506  0.158 -0.252
object     -0.439 -0.038 -0.856  0.502 -0.668 -0.439  0.066
            parang block  object
info
comp
arith
simil
digit
vocab
pictcomp
parang      0.000
block       0.369  0.000
object     -0.031  0.104  0.000
```

```
$mean
       info     comp    arith    simil    digit    vocab
          0        0        0        0        0        0
    pictcomp   parang    block   object
          0        0        0        0
```

The first part of the output is the covariance matrices and the second part
is the means of the manifest variables. The means are 0 for both functions
because cfa does not estimate the means by default. For the means returned
from the fitted function, the 0s would be replaced with the indicator means
if meanstructure = TRUE was used in the cfa function. This update would

not affect the output returned from the `residual` function (i.e., the means would still be 0).

4.4.2   CFA with Categorical Data

In the last section, we demonstrate how to perform a CFA with categorical data. There are two ways to set up the data for **lavaan** if it is categorical:

1. If an indicator is nominal, we must create C - 1 dummy variables (where C is the number of categories of the indicator). This is akin to how we would handle it in a traditional regression problem, although many functions in R, such as the `lm` function, automatically create the dummy variables for us.

2. If an indicator is ordinal, we can tell R that the indicator is ordinal with the `ordered` function prior to running `cfa`. Alternatively, we can pass the `ordered` argument when we fit the model using the `cfa` function.

By default, when the data are categorical, **lavaan** will switch to the weighted least squares estimator, which uses diagonally weighted least squares. We will illustrate the two approaches mentioned above by recoding the `wiscsem` data set as ordinal variables. Please note this is strictly done for didactic purposes and not something readers should ordinarily do with their data.

We save `wiscsem` as `wiscsem_cat` and apply a function (`quart_cut`), which splits an indicator into quartiles and assigns each observation to a quartile via the `lapply` function. The `lapply` function allows us to apply the `quart_cut` function to all the indicators simultaneously rather than individually.

```
wiscsem_cat <- wiscsem
wiscsem_cat[ ,] <- lapply(wiscsem_cat[ ,], quart_cut)
```

4.4.2.1   Ordinal CFA–Method 1

Method 1 involves telling R the indicators are ordinal prior to using the `cfa` function. Right now the indicators are just numeric values of 1, 2, 3, or 4 corresponding to which quartile an observation fell into. We can convert these numeric values to an ordinal scale again using the `lapply` command. First, we save `wiscsem_cat` as `wiscsem_ord`. Then we tell `lapply` to convert all the variables in `wiscsem_ord` to ordinal variables using the `ordered` function and save it as `wiscsem_ord` again.

```
wiscsem_ord <- wiscsem_cat
wiscsem_ord[,] <- lapply(wiscsem_ord[ ,], ordered)
```

Next, we fit our CFA model using the wiscsem_ord dataset instead of wiscsem. We save the results from this model as iq_fit_ord1 and then print the output using the summary function.

```
iq_fit_ord1 <- cfa(iq_mod, wiscsem_ord)
summary(iq_fit_ord1)
```

```
lavaan (0.5-23.1097) converged normally after  21 iterations

  Number of observations                              175

  Estimator                                DWLS        Robust
  Minimum Function Test Statistic        42.611        60.600
  Degrees of freedom                         43            43
  P-value (Chi-square)                    0.488         0.039
  Scaling correction factor                             0.808
  Shift parameter                                       7.882
    for simple second-order correction (Mplus variant)

Thresholds:
              Estimate   Std.Err   z-value   P(>|z|)
    info|t1     -0.725     0.105    -6.925     0.000
    info|t2     -0.343     0.097    -3.536     0.000
    info|t3      0.670     0.103     6.493     0.000

Scales y*:
              Estimate   Std.Err   z-value   P(>|z|)
    info         1.000
```

The output again is abridged and we print only some of the model output above. When we run this command, we get a warning message that indicates zero cells in the rows or columns of the bivariate table for some of the indicators. We receive this warning message because **lavaan** is calculating polychoric correlations. We can ignore the warning message here and review the output returned from the summary function. In the output, we now have an additional column of robust estimators of model fit. We strongly recommend using this sandwich estimator (Asparouhov & Muthén, 2010), which has been discussed extensively elsewhere.[3] The summary output also includes information on the indicator thresholds and scales (shown above just for the variable "info"). Otherwise, the output is similar to the usual call to cfa.

---

[3]https://groups.google.com/forum/#!topic/lavaan/wYA9msIv5TI

4.4.2.2 Ordinal CFA–Method 2

This method involves telling the `cfa` function that the indicators are ordinal. This additional adjustment can be made as follows:

```
iq_fit_ord2 <- cfa(iq_mod, wiscsem_cat,
                   ordered = names(wiscsem_cat))
```

We specify `ordered = names(wiscsem_cat)` to tell `cfa` to change all the indicators in `wiscsem_cat` to be ordered. The summary output from this call is identical to `iq_fit_ord1`. If two of the indicators were ordered, then we would specify only those two indicators (e.g., if verb and info were ordinal but nothing else).

```
iq_fit_ord2 <- cfa(iq_mod, wiscsem_cat,
                   ordered = c("verb", "info"))
```

4.5 Summary

In this chapter, we provided a brief introduction to exploratory and confirmatory factor analyses and demonstrated how to perform EFA using `factanal` and the **psych** package and CFA using the **lavaan** package. We began with a brief review of the common factor model and then moved into examples using EFA and CFA. We recommend that our readers review other resources on factor analysis and SEM (e.g., Kline, 2015) to have a better understanding of these methodologies. Next, we want to note that in addition **lavaan**, there are a great number of R packages that are capable of estimating factor analytic and structural equation models, such as **sem** (Fox, Nie, & Byrnes, 2017), **FactoMineR** (Lê, Josse, & Husson, 2008), and **OpenMx** (Neale et al., 2016). In the next chapter, we cover item response theory models, which typically assume a single underlying latent variable. A unidimensional solution, found via factor analysis, is an important assumption of the models presented in Chapters 5 and 6.

5

Item Response Theory for Dichotomous Items

5.1 Chapter Overview

This chapter introduces unidimensional item response theory (IRT) models for dichotomously scored items. The chapter begins with a brief description of the IRT framework. The remainder of the chapter proceeds as follows: Unidimensional IRT models are briefly introduced, with their mathematical formulation; parameters are then described in order to facilitate interpretations; and finally, R examples showing how to estimate unidimensional IRT models for dichotomous items and extract information are presented. The IRT models covered in this chapter include the Rasch model, 1-parameter, 2-parameter, 3-parameter, and 4-parameter IRT models. Several user-contributed packages exist on CRAN to fit IRT models, such as **eRm** (Mair & Hatzinger, 2007a), **ltm** (Rizopoulos, 2006), **mirt** (Chalmers, 2012), and **TAM** (Kiefer, Robitzsch, & Wu, 2016). In this chapter, we use the **mirt** package (Chalmers, 2012) to demonstrate how to estimate unidimensional IRT models because it is the most comprehensive and up-to-date R package for estimating various IRT models.

5.2 Introduction

5.2.1 Comparison to Classical Test Theory

In Chapter 2, the classical test theory (CTT) model was introduced. IRT was developed to overcome the shortcomings of CTT model. Specifically, IRT differs from CTT by making several theoretical and testable assumptions (see Embretson & Reise, 2000, Chapter 1). First, IRT is a probabilistic model of how examinees respond to any given item(s) based on their level of the underlying latent trait. The probability of responding to an item correctly, or more generally endorsing an item, is a monotonically increasing function of the measured latent trait (denoted by θ). This function may depend on the difficulty (or location), discrimination (or lower), and guessing (i.e., lower asymptote) parameters of the item. Unlike IRT, CTT focuses solely on observable (i.e., raw) scores rather than the relationship between the items and the latent trait being measured.

Second, the latent trait estimates obtained from the IRT models are assumed to be independent of items on the test and the characteristics of the examinee population, whereas the test scores obtained from CTT heavily depend on the selected items on the test. Even if the difficulty levels of the items vary from one test to another, the latent trait estimates from IRT are expected to be very similar, whereas the test scores from CTT can differ significantly across the tests. Therefore, it is much easier to compare the examinees' performances between different tests within the IRT framework compared to the CTT framework.

Third, IRT assumes that the item parameters are invariant across subgroups of examinees and across multiple test administrations. As a result of this assumption, the items are expected to have the same parameters regardless of which subgroup the examinees come from (e.g., examinees of different gender or ethnicity).[1] Also, the items across multiple test administrations are considered mutually invariant if the item parameters can be linearly transformed (Hambleton, Swaminathan, & Rogers, 1991; Rupp & Zumbo, 2006; Stocking & Lord, 1983). The parameter invariance property in IRT allows us to solve important measurement problems that were difficult to handle in CTT, such as those encountered in test equating and computerized adaptive testing (Hambleton et al., 1991).

5.2.2   Basic Concepts in IRT

Within the IRT framework, the items are characterized individually and the test characteristics are derived from the items. These characteristics include difficulty, the latent trait level required for answering an item correctly,[2] discrimination, power of an item to distinguish individuals with low and high latent trait levels, and guessing, the probability of answering an item correctly for the examinees whose latent trait levels are very low. Based on the difficulty, discrimination, and guessing levels of the item, we can draw an item

---

[1]This is known as measurement invariance or differential item functioning. These topics are presented in Chapter 11.

[2]In education, we typically talk about responding to an item correctly. In other settings, such as psychology or medicine where one may be responding to a survey without a correct answer, one thinks, more generally, about the probability of endorsing an item. Therefore, a reader coming from these fields can safely substitute that wording.

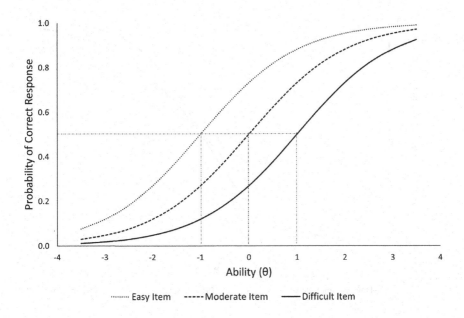

FIGURE 5.1
Item characteristic curves of the items with easy, moderate, and high difficulty.

characteristic curve (ICC) that shows an examinee's probability of correctly answering the item conditional on his or her latent trait (θ) level.

Figure 5.1 shows the ICCs of three items that have the same discrimination and guessing levels but different levels of difficulty (i.e., easy, moderate, or difficult). To have a probability of 0.5 of answering the item correctly, the required level of the latent trait, θ, is -1 for the easy item, 0 for the moderate item, and 1 for the difficult item. Because the ICC is a logistic probability curve, the latent trait is on a logistic scale, which usually ranges from -5 to +5. The same logistic scale is also used for determining the difficulty of the item.

Figure 5.2 shows the ICCs of three items that have the same difficulty and guessing levels but different levels of discrimination (i.e., low, moderate, or high). Item discrimination is essentially the slope of the ICC at the location (difficulty) of the item (i.e., $\theta = 0$ in Figure 5.2). The steeper the ICC, the more discriminating the item is between low and high ability examinees. In Figure 5.2, the ICC of the item with the lowest discrimination (the dotted line) has the flattest slope. As the slopes increase for the other two items, the discriminating power of the items improves and these items are more discriminating at their item locations.

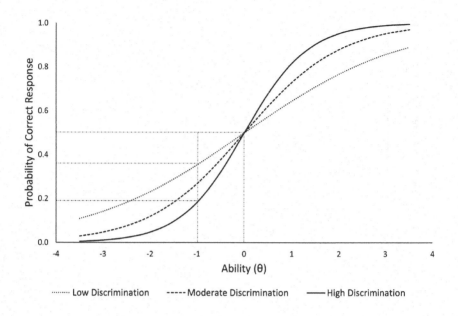

FIGURE 5.2
Item characteristic curves of the items with low, moderate, and high discrimination.

Figure 5.3 shows the ICCs of two items that have the same difficulty and discrimination levels but different levels of guessing (i.e., low or high). The guessing level increases the intercept of the ICC (i.e., the point where ICC intersects with the y axis). The higher the guessing, the more likely the examinees with a low latent trait level can answer the item correctly. In Figure 5.3, the ICC of the item with the high guessing (the dotted line) has a larger intercept that enables an examinee with a low latent trait level (e.g., $\theta = -1$) to have nearly a 60% chance of answering the item correctly. For the other item with a low guessing level (i.e., solid line), the same examinee has nearly a 35% chance of answering the item correctly. The impact of the guessing level decreases as the latent trait level increases.

Another important concept in IRT is the item information function (IIF). IIF indicates the amount of information for an item along the latent trait continuum. For a fixed level of the latent trait, the greater the IIF is, the more information is available for examinees at that particular level of the latent trait. The information function for item i ($i = 1, 2, \ldots, N$) at the latent trait level θ is denoted as $I_i(\theta)$. Because items are assumed to be locally independent (see below), we can sum the IIFs for all items and calculate the test information function (TIF) as follows:

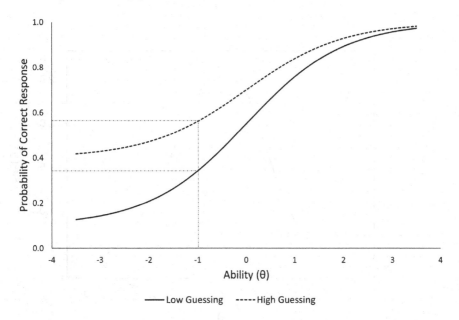

FIGURE 5.3
Item characteristic curves of the items with low and high guessing.

$$I(\theta) = \sum_{i=1}^{N} I_i(\theta), \qquad (5.1)$$

where $I(\theta)$ is the total test information for the latent trait θ. Using the TIF, we can compute the conditional standard error of measurement (cSEM), which provides information about the precision of the test at a given latent trait level. The cSEM value is the square root of the reciprocal value of the TIF at θ:

$$cSEM(\theta) = \sqrt{\frac{1}{I(\theta)}}. \qquad (5.2)$$

Figure 5.4 shows the reciprocal relationship between the TIF (solid line) and cSEM (dashed line). The maximum point of the TIF corresponds to the lowest point of the cSEM. In other words, where the information on a test or item is at a maximum, we have greater precision for estimating the level of the latent trait for examinees around this point. In Figure 5.4, our test is most precise at estimating the latent trait for examinees with θ near 0.

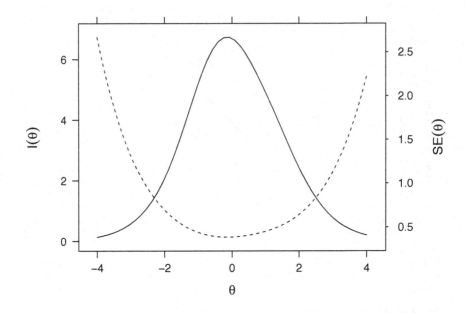

FIGURE 5.4
The reciprocal relationship between TIF (the solid line) and cSEM (the dashed line).

An important corollary of this is that examinees on a test will have different cSEMs depending on their latent trait levels.

5.2.3   IRT Model Assumptions

There are two major modeling assumptions in IRT. The are unidimensionality and local independence. The unidimensionality assumption requires that there is a single latent trait underlying a set of test items. Finding a dominant component or factor affecting test performance is required to meet this assumption. The methods for checking the unidimensionality assumption include principal component analysis, exploratory factor analysis, confirmatory factor analysis, DIMTEST (Strout, 1990), and DETECT (Zhang & Stout, 1999). Of these methods, we discussed exploratory and confirmatory factor analyses in Chapter 4 (the most common methods for assessing dimensionality).

The local independence assumption requires the probability of successfully responding to a given item to be based soley on the latent trait, regardless of the responses given to other items on the test. In other words, when we condition on the latent trait level, local independence implies that no relationship remains between the items (Embretson & Reise, 2000). When the assumption of unidimensionality holds, the local independence assumption is also met because the two concepts are related to each other only through the single latent trait (Lord, 1980). Yen's Q3 statistic (Yen, 1984) is one of the most commonly used methods for checking the local independence assumption.

In addition to unidimensionality and local independence of test items, there are several assumptions that should be checked before estimating item parameters and latent trait levels. The non-speeded test administration assumption requires all examinees to have adequate time to answer the items on the test. In the Rasch, one-, and two-parameter IRT models, guessing the correct answer to a given item is assumed to have minimal or no effect on the probability of answering the item correctly. Based on this assumption, the ICCs of the one- and two-parameter models have a lower asymptote of zero. This assumption indicates that examinees with a low latent trait level would have a very small probability of answering the item correctly. The final assumption is that the item discrimination is equivalent across the items on the test, which is only assumed for the Rasch and one-parameter IRT models. This assumption constrains the item discrimination level to be the same (i.e., "1" or another estimated value) across all items.

5.3 The Unidimensional IRT Models for Dichotomous Items

There are four unidimensional IRT models for dichotomously scored test items:[3] the one-parameter, two-parameter, three-parameter, and four-parameter IRT models. These models differ based on their assumptions, and therefore modeling constraints, regarding discrimination and guessing.

5.3.1 One-Parameter Logistic Model and Rasch Model

5.3.1.1 One-Parameter Logistic Model

The simplest IRT model is the one-parameter logistic IRT model (also known as the 1PL model). Mathematically, the 1PL model can be written as

---

[3]Technically, there are five, but the Rasch model is considered a special case of the one-parameter IRT model.

$$P(Y_{ij} = 1|\theta_j, a, b_j) = \frac{\exp(Da(\theta_j - b_i))}{1 + \exp(Da(\theta_j - b_i))}, \qquad (5.3)$$

where θ_j is the level of the latent trait of person j ($j = 1, ..., J$), a is the item discrimination parameter, b_i is the item difficulty parameter for item i ($i = 1, ..., I$), D is a scaling constant to place the parameters of the logistic model onto the scale of a normal ogive model (when $D = 1.7$). This equation states that the probability of person j responding correctly to item i is a function of the latent trait, the item discrimination, and the difficulty of the item. Notice in Equation 5.3 that there is no subscript on item discrimination and no mention of guessing. This indicates that the 1PL model assumes that the items have the same discrimination and that there is no effect of guessing on responding correctly to an item.

To demonstrate how to estimate the 1PL model in R, we again consider the SAPA data set contained in the **hemp** package, which we introduced in Chapter 2. To estimate the item parameters, we use the mirt function from the **mirt** package (Chalmers, 2012). In order to make sure the functions in the **mirt** package are available, we should first install the **mirt** package. Then, we activate both **mirt** and **hemp** using the library command.

```
install.packages("mirt")
library("mirt")
library("hemp")
```

To define the 1PL model in mirt for the SAPA data set, we use **mirt**'s model syntax and save it as onepl_mod.

```
onepl_mod <- "F = 1 - 16
              CONSTRAIN = (1 - 16, a1)"
```

The first line of this code states that a single latent trait, F, is manifested by the items in columns 1 through 16 in the SAPA data set. The second line, beginning with CONSTRAIN, constrains the items in columns 1 through 16 to have the same item discrimination. The **mirt** package labels the item discrimination parameter for a unidimensional IRT model as a1, not a as shown in Equation 5.3. If we wanted to fit a 1PL to just items in columns 1 through 4, then we would change 1 - 16 to 1 - 4. If we wanted to use the items in columns 1 through 3, 5, and 9 though 16, we would define it as 1 - 3, 5, 9 - 16.

The syntax of the mirt package can be a little cumbersome and difficult at first. However, we highly recommend learning this package for estimating IRT models due to its capability to estimate various IRT models (e.g., polytomous IRT models, multidimensional IRT models, and explanatory IRT models[4]).

---

[4]We will talk about these models in detail in the following chapters.

Next, we fit the 1PL model using the model defined above and instruct the `mirt` function to estimate standard errors for the item parameters by setting `SE = TRUE`. We save the fitted model to `onepl_fit`.

```
onepl_fit <- mirt(data = SAPA, model = onepl_mod, SE = TRUE)
onepl_params <- coef(onepl_fit, IRTpars = TRUE, simplify = TRUE)
```

The `onepl_fit` object contains the estimated item parameters, the mean of the latent trait, the variance-covariance matrix of the latent trait, and additional information regarding the estimation process. After fitting the model, we extract the estimated parameters and the mean and covariance matrix of the latent trait with the `coef` function and save it to `onepl_params`. During this step, we used `IRTpars = TRUE` to convert the slope and intercept parameters (the default parameterization in `mirt`) into the traditional IRT parameters and use `simplify = TRUE`, which combines the item parameters into a single data frame rather than a long list. The transformation of the intercept parameter into the traditional item difficulty parameter can be accomplished using the following formula:

$$b_i = \frac{-d_i}{a1_i}, \tag{5.4}$$

where d_i is the intercept parameter, $a1_i$ is the slope (i.e., discrimination) parameter, and b_i is the traditional item difficulty parameter. The main difference between d_i and b_i is that b_i represents item difficulty whereas d_i represents item easiness.

Finally, to print the estimated item parameters in a more compact way, we save the data frame containing the items to `onepl_items` and view the first few rows with the `head` function.

```
onepl_items <- onepl_params$items
head(onepl_items)
```

	a	b	g	u
reason.4	1.445587	-0.5557199	0	1
reason.16	1.445587	-0.8020747	0	1
reason.17	1.445587	-0.7980649	0	1
reason.19	1.445587	-0.4546611	0	1
letter.7	1.445587	-0.3923381	0	1
letter.33	1.445587	-0.2810892	0	1

Each row starts with the item name and the columns correspond to the estimated item parameters. The first column (a) is item discrimination, the second column (b) is item difficulty, the third column (g) is the lower asymptote (i.e., guessing), and the last column (u) is the upper asymptote. Because the 1PL model does not include the lower and upper asymptote parameters, we will ignore the last two columns as they will always be 0 and 1, respectively. (The details of the lower and upper asymptote parameters will be explained later in the section on the 3PL and 4PL models.) The first column shows the estimated item discrimination parameter of 1.446. Because we used the 1PL model, the item discrimination parameter is the same across all items. The second column shows the estimated item difficulty parameters. In the SAPA data set, the second item (reason.16) is the easiest item and the last item (rotate.8) is the most difficult item on the test (not shown in the printed output above).

To see the standard errors for the estimated item parameters, we need to re-run the coef function with the printSE = TRUE argument. We do this below and print the contents of the output with the name function.

```
onepl_se <- coef(onepl_fit, printSE = TRUE)
names(onepl_se)
```

```
 [1] "reason.4"  "reason.16" "reason.17" "reason.19"
 [5] "letter.7"  "letter.33" "letter.34" "letter.58"
 [9] "matrix.45" "matrix.46" "matrix.47" "matrix.55"
[13] "rotate.3"  "rotate.4"  "rotate.6"  "rotate.8"
[17] "GroupPars"
```

We can see that the object onepl_se is a list that contains information about each item separately. To see the information for the most difficult item, rotate.8, we can extract it using the $ symbol as shown below:

```
onepl_se$rotate.8
            a1            d logit(g) logit(u)
par 1.44558726 -2.00517306     -999      999
SE  0.03612038  0.08517212       NA       NA
```

The first row corresponds to the estimated parameters and the second row to the standard errors. The first column again corresponds to item discrimination but the second column no longer corresponds to the item difficulty (note that b is now d). For readers requiring standard errors for item difficulty, the ltm package (Rizopoulos, 2006) can also provide this information. The -999, 999, and NAs are because the lower and upper asymptote parameters were not estimated for the model.

The **mirt** package provides two functions, plot and itemplot, that can plot various item and test characteristics. These are actually just wrapper functions for the xyplot function in the **lattice** package. So, if we wanted to customize our plots, we could use any graphics argument (e.g., changing the color or line type) available for the xyplot function (see ?xyplot for more details).

The example below shows how to create ICCs for the first two items. In

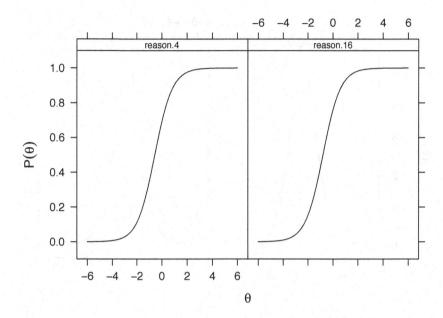

FIGURE 5.5
Item characteristic curves for items 1 (reason.4) and 2 (reason.16) for the 1PL model.

Figure 5.5, the x-axis in the ICCs shows the latent trait (θ), which in the case of the SAPA data set could be considered as either intelligence or aptitude, and the y-axis in the ICCs shows the probability of answering the item correctly, $P(\theta)$. Because the 1PL model constrains the discrimination parameters to be the same across all items, the slopes of the ICCs are identical but just shifted along the x-axis based on the item difficulty parameters.

```
plot(onepl_fit, type = "trace", which.items = 1:2)
```

We want to remind our readers that once they run the code presented above, their plots will be in color, while ours are printed in grayscale here.

An alternative way to create ICCs is to use the itemplot function. In the following example, we use the itemplot function for the first two items again.

```
itemplot(onepl_fit, type = "trace", item = 1)
itemplot(onepl_fit, type = "trace", item = 2)
```

An item information plot, which shows the amount of item information as a function of the latent trait, can also be easily created using the plot

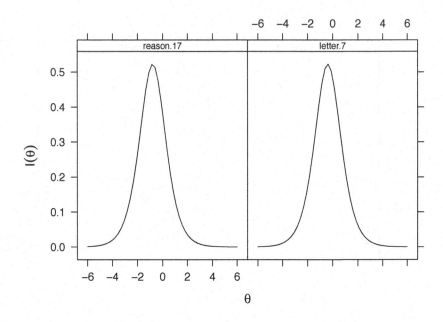

FIGURE 5.6
Item information functions for items 3 (reason.17) and 5 (letter.7) for the 1PL model.

function. Below, we show how to plot the IIFs for the third and fifth items in the `SAPA` data set (see Figure 5.6).

```
plot(onepl_fit, type = "infotrace", which.items = c(3, 5))
```

In both Figure 5.5 and Figure 5.6, the ICCs and IIFs are plotted separately for the items. It is possible to combine the plots for the two items within a single plot using the `facet_items = FALSE` argument. In the following example, we combine the ICC plots for items 1 and 2 together. We also set the `auto.key = list(points = FALSE, lines = TRUE, columns = 2)` and `par.settings = simpleTheme(lty = 1:2)` to separate line types for each ICC and to draw a two-column legend indicating the items. Figure 5.7 shows the combined ICCs for item 1 (solid line) and item 2 (dashed line).

```
plot(onepl_fit, type = "trace", which.items = 1:2,
     facet_items = FALSE,
     auto.key = list(points = FALSE, lines = TRUE, columns = 2),
     par.settings = simpleTheme(lty = 1:2))
```

In addition to ICCs and IIFs, we can also create a TIF plot and a cSEM

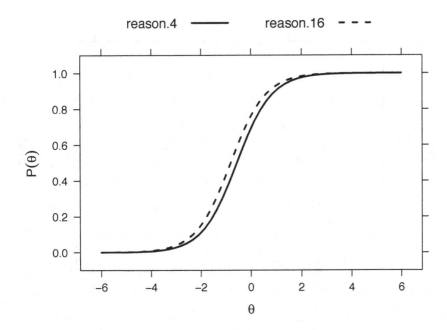

FIGURE 5.7
Combined ICCs for items 1 (reason.4) and 2 (reason.16) for the 1PL model.

plot. In the following example, we specify a range of abilities (-3 to 3) and show how to create a combined TIF and cSEM plot for the 1PL **SAPA** model.

```
plot(onepl_fit, type = "infoSE", theta_lim = c(-3, 3))
```

Figure 5.8 shows that the items in the **SAPA** data set are highly informative and contain the least amount of measurement error between $\theta = -1$ and $\theta = 1$. We can also create separate TIF and cSEM plots using the **type = "info"** and **type = "SE"** arguments in the **plot** function.

```
plot(onepl_fit, type = "info", theta_lim = c(-3, 3))
plot(onepl_fit, type = "SE", theta_lim = c(-3, 3))
```

5.3.1.2 Rasch Model

The Rasch model (Rasch, 1960) is a specific form of the 1PL model where the item discrimination is set to 1 for all items. As a consequence of this adjustment, the probability of answering an item correctly is a function of only the item difficulty and latent trait. Mathematically, the Rasch model can be written as:

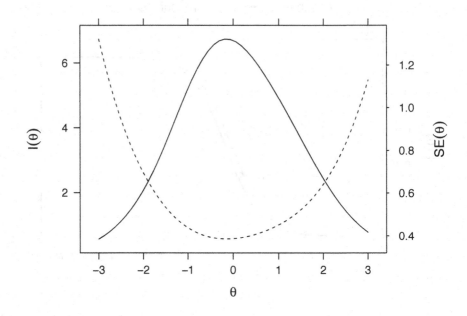

FIGURE 5.8
The TIF and cSEM plot for the 1PL model.

$$P(Y_{ij} = 1|\theta_j, b_i) = \frac{\exp(D(\theta_j - b_i))}{1 + \exp(D(\theta_j - b_i))}, \tag{5.5}$$

where all the components are the same as the 1PL model except the common item discrimination parameter, which can be omitted because $a = 1$ for all items.[5]

Because the item discrimination parameter has no effect on the probability of correctly answering an item, a simple explanation can be used to describe the relationship between item difficulty and latent trait. If an examinee's latent trait level matches the difficulty of the item, then the examinee has a 50% chance of answering the item correctly (assuming no guessing is involved). Under the Rasch model, examinees are ordered the same in terms of predicted responses regardless of which item difficulty location is used. Similarly, items are ordered the same in terms of predicted responses regardless of what level of the latent trait is considered.

To illustrate the Rasch model, we again use the SAPA data set. Unlike the

---

[5]Note, we derive the Rasch model from Equation 5.3 by setting a to 1.

1PL model, the Rasch model syntax does not require the CONSTRAIN command. Instead, we specify itemtype = "Rasch" in the mirt function to estimate the item parameters according to the Rasch model and save the fitted model to rasch_fit.

```
rasch_mod <- "F = 1 - 16"
rasch_fit <- mirt(data = SAPA, model = rasch_mod,
                  itemtype = "Rasch", SE = TRUE)
```

As we have done for the 1PL model earlier, we can extract the estimated item parameters using the coef function, save them (rasch_params), extract the item parameters as a data frame (rasch_items), and finally print the first six rows using the head function.

```
rasch_params <- coef(rasch_fit, IRTpars = TRUE, simplify = TRUE)
rasch_items <- rasch_params$items
head(rasch_items)
```

```
          a          b g u
reason.4  1 -0.8024929 0 1
reason.16 1 -1.1585935 0 1
reason.17 1 -1.1528004 0 1
reason.19 1 -0.6564121 0 1
letter.7  1 -0.5663314 0 1
letter.33 1 -0.4055267 0 1
```

In the output, the first column (i.e., item discrimination) is 1 for all the items in the Rasch model, whereas the item discrimination was estimated as 1.446 for all the items in the 1PL model. In addition to item discrimination, the second column indicates that the item difficulty parameters from the Rasch model are different from the item difficulty parameters in the 1PL model.

As with the 1PL, we can plot the ICCs for the Rasch model. Based on the results from the Rasch model, item 2 (reason.16) was the easiest item and item 16 (rotate.8) was the most difficult item in the SAPA data set. These are the same items that the 1PL model identified as the easiest and hardest items. Below, we plot the ICCs for these two items (see Figure 5.9).

```
plot(rasch_fit, type = "trace", which.items = c(2, 16))
```

Note that the Rasch model assumes that the item discrimination parameter is equal to 1 for all items, and therefore, the ICCs are less steep in Figure 5.9 than those presented in Figure 5.5. Figure 5.9 shows that even the examinees with a low ability on the latent trait have a high probability of answering item 2 correctly, whereas only the examinees with a high ability, of at least $\theta = 2$, have a 50% probability of answering item 16 correctly. IIFs and TIFs can also be created in a similar manner for the Rasch model as we have demonstrated for the 1PL model. We leave this as an exercise for our readers.

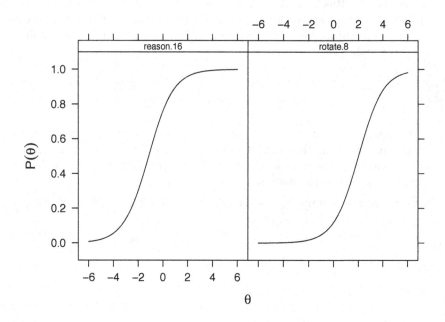

FIGURE 5.9
Item characteristic curves for items 2 (reason.16) and 16 (rotate.8) for the Rasch model.

5.3.2 Two-Parameter Logistic Model

The two-parameter logistic (2PL) model is a more flexible form of the 1PL and Rasch models. The 2PL model allows for each item to have its own item discrimination parameter. The mathematical formulation of the 2PL model is as follows:

$$P(Y_{ij} = 1|\theta_j, a_i, b_i) = \frac{\exp(Da_i(\theta_j - b_i))}{1 + \exp(Da_i(\theta_j - b_i))}, \tag{5.6}$$

where the parameters are the same as described above for Equation 5.3 with the exception that a_i is a unique discrimination parameter for item i.

In the following example, we estimate the item parameters for the SAPA data set using the 2PL model. The model syntax for estimating the 2PL model is the same way as it was for the Rasch model. The only difference comes when we call the mirt function. This time we need to pass the itemtype = "2PL" argument to estimate the 2PL model. As before, we save the fitted model (twopl_fit), extract the model parameters (twopl_params), and print the item parameters for all the items.

```
twopl_mod <- "F = 1 - 16"
twopl_fit <- mirt(data = SAPA, model = twopl_mod,
                  itemtype = "2PL", SE = TRUE)
twopl_params <- coef(twopl_fit, IRTpars = TRUE, simplify = TRUE)
twopl_items <- twopl_params$items
twopl_items
```

	a	b	g	u
reason.4	1.6924256	-0.5127258	0	1
reason.16	1.4616058	-0.7967194	0	1
reason.17	1.8568189	-0.7052519	0	1
reason.19	1.4429276	-0.4544282	0	1
letter.7	1.5739581	-0.3749607	0	1
letter.33	1.3512472	-0.2906578	0	1
letter.34	1.6568903	-0.4165187	0	1
letter.58	1.4637541	0.2090402	0	1
matrix.45	1.0649705	-0.1241352	0	1
matrix.46	1.1060157	-0.2292152	0	1
matrix.47	1.3463316	-0.4666121	0	1
matrix.55	0.8786048	0.6793708	0	1
rotate.3	1.7878172	1.1986461	0	1
rotate.4	2.0841977	1.0317428	0	1
rotate.6	1.6388551	0.7524753	0	1
rotate.8	1.5855260	1.3201267	0	1

In the output from the 2PL model, both item discrimination (first column) and item difficulty (second column) parameters are different for each item. Item 14 (rotate.4) has the highest item discrimination (2.084), whereas item 12 (matrix.55) has the lowest item discrimination. Next, we visually examine the differences between these two items by plotting their ICCs using the plot function.

```
plot(twopl_fit, type = "trace", which.items = c(12, 14))
```

Figure 5.10 shows that the slope of rotate.4 (on the right) is much steeper than the slope of matrix.55 (on the left) because of the difference in the item discrimination parameters. This difference suggests that when the 2PL model is used for estimating item parameters, item 14 (rotate.4) can distinguish examinees with low and high latent traits better than item 12 (matrix.55).

We can also compare the ICCs for items with different difficulty parameters but similar discriminations. For example, in the SAPA data set, item 5 (letter.7) and item 16 (rotate.8) have similar item discrimination parameters but the items differ in terms of difficulty. Figure 5.11, created using the code shown below, shows that at an estimated latent trait level of $\theta = 0$, the prob-

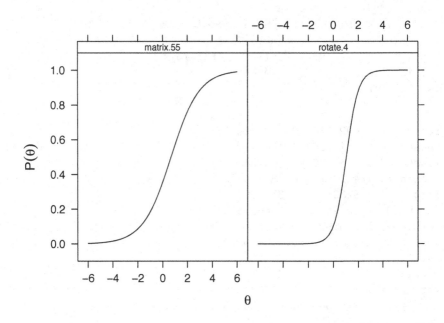

FIGURE 5.10
Item characteristic curves for items 12 (matrix.55) and 14 (rotate.4) for the 2PL model.

ability of answering letter.7 (the easier item) correctly is nearly .7, whereas the probability of answering rotate.8 (the harder item) correctly is only .1.

```
plot(twopl_fit, type = "trace", which.items = c(5, 16),
    facet_items = FALSE, auto.key = list(points = FALSE,
    lines = TRUE, columns = 2),
    par.settings = simpleTheme(lty = 1:2))
```

5.3.3 Three-Parameter Logistic Model

The three-parameter logistic (3PL) model is just an extension of the 2PL model. This model includes the same item parameters (i.e., a_i and b_i) from the 2PL model and an additional parameter (c_i) that represents the likelihood of endorsing the item due solely to chance or guessing. The c_i parameter is known as either the lower asymptote or the pseudo guessing parameter. The 3PL model can be written as:

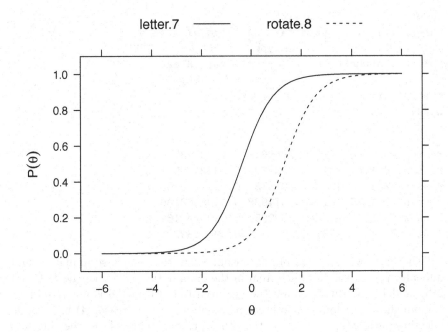

FIGURE 5.11
Item characteristic curves for items 5 (letter.7) and 16 (rotate.8) for the 2PL model.

$$P(Y_{ij} = 1|\theta_j, a_i, b_i, c_i) = c_i + \frac{1 - c_i}{1 + \exp(-Da_i(\theta_j - b_i))}. \qquad (5.7)$$

Unlike in the Rasch, 1PL, and 2PL models, the probability of a correct response in the 3PL model does not approach zero as the latent trait goes to $-\infty$. Instead, the probability approaches a positive value (usually $1/k$, where k is the number of response categories in a multiple-choice item). The inclusion of the c_i parameter is due to the fact that some examinees with a very low latent trait are likely to select a response option randomly, and thus random guessing would result in selecting the correct response with a probability equal to c_i.

In the following example, we estimate item parameters for the SAPA data set using the 3PL model. When we fit the model, we specify the itemtype = "3PL" argument to estimate the 3PL model. Below we define, fit, and save the fitted model, extract the parameters, and print the first few rows of the item parameters.

```
threepl_mod <- "F = 1-16"
threepl_fit <- mirt(data = SAPA, model = threepl_mod,
                    itemtype = "3PL", SE = TRUE)
threepl_params <- coef(threepl_fit, IRTpars = TRUE,
                       simplify = TRUE)
threepl_items <- threepl_params$items
head(threepl_items)
```

```
                 a           b            g u
reason.4   1.965832  -0.2869613  0.1220243277 1
reason.16  1.419231  -0.7995733  0.0022155571 1
reason.17  1.780427  -0.7076754  0.0020240288 1
reason.19  1.379131  -0.4514928  0.0005308146 1
letter.7   1.535846  -0.3622950  0.0007297169 1
letter.33  1.336533  -0.2751526  0.0010139285 1
```

The first two columns of the output show the item discrimination and difficulty parameters for each item in the SAPA data set. Unlike the 1PL and 2PL models, the third column (i.e., g) from the 3PL model is not zero. This column shows the lower asymptote parameters. Item 1 (reason.4) has a large guessing parameter relative to the other printed items where guessing was typically minimal.

To see the effect of guessing on the items, we draw the ICC plot for items 1 (reason.4) and 4 (reason.19). Figure 5.12 shows that as the latent trait gets smaller (e.g., $\theta < -2$), the probability of answering item 1 correctly approaches .12, while the probability is near zero for item 4.

```
plot(threepl_fit, type = "trace", which.items = c(1, 4),
     facet_items = FALSE, auto.key = list(points = FALSE,
     lines = TRUE, columns = 2),
     par.settings = simpleTheme(lty = 1:2))
```

5.3.4 Four-Parameter Logistic Model

The four-parameter logistic (4PL) model, introduced by Barton and Lord (1981), is a special case of the 3PL model. The 4PL model allows for items to have both lower and upper asymptote parameters. The upper asymptote parameter can be described as a ceiling parameter that prevents the probability of responding correctly from ever approaching 1 (i.e., 100%) regardless of the examinee's latent trait level. The 4PL is written as:

$$P(Y_{ij} = 1|\theta_j, a_i, b_i, c_i, u_i) = c_i + \frac{u_i - c_i}{1 + \exp(-Da_i(\theta - b_i))}, \qquad (5.8)$$

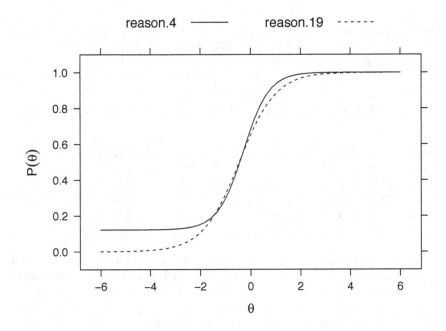

FIGURE 5.12
Item characteristic curves for reason.4 (item 1) and reason.19 (item 4) for the 3PL model.

where the u_i parameter represents the upper asymptote. This additional parameter makes the numerator in the equation less than 1 when the c_i parameter is not equal to 0 and the u_i parameter is not equal to 1 (see Magis (2013) for a detailed discussion of the 4PL model).

Although the 3PL model is more common for modeling latent traits such as achievement and aptitude, the 4PL model can be more suitable for non-cognitive personality traits — such as neuroticism, openness, aggression, and motivation — because the probability of reaching the highest level of such traits may not approach 1. The benefit of including an upper asymptote parameter in the IRT model has been demonstrated in several studies (Loken & Rulison, 2010; Osgood, McMorris, & Potenza, 2002; Reise & Waller, 2009; Tavares, Andrade, & Pereira, 2004).

In the following example, we estimate the 4PL model for the SAPA data set. We set itemtype = "4PL" to estimate the 4PL model. Like before, we save the fitted model as fourpl_fit and then extract the estimated item parameters and print the first few rows of the item parameters.

```
fourpl_mod <- "F = 1 - 16"
fourpl_fit <- mirt(data = SAPA, model = fourpl_mod,
                   itemtype = "4PL", SE = TRUE)
fourpl_params <- coef(fourpl_fit, IRTpars = TRUE,
                      simplify = TRUE)
fourpl_items <- fourpl_params$items
head(fourpl_items)
```

	a	b	g	u
reason.4	1.976659	-0.3388019	0.1106491560	0.9912226
reason.16	1.604933	-0.8563556	0.0003777105	0.9670460
reason.17	2.064633	-0.7469289	0.0006907810	0.9718755
reason.19	1.912908	-0.6232106	0.0004655221	0.9056812
letter.7	3.598660	-0.3505197	0.1257669007	0.8856442
letter.33	3.400545	-0.3093899	0.1338977317	0.8529766

In the output, we see that the last column (i.e., upper asymptote) is not equal to zero this time because we used the 4PL model for the estimation. Although the upper asymptote parameter is close to 1 for most of the items, there are several items that have a lower upper asymptote (e.g., letter.7, letter.33). The maximum probability of responding to these items correctly does not approach 1 even for examinees with high levels of the latent trait. As we have done for the previous models, we can also visually examine the effect of the item parameters from the 4PL model on the ICCs. The following code shows the commands for creating the ICC plots for items 13 and 16.

```
plot(fourpl_fit, type = "trace", which.items = c(13, 16))
```

5.4   Ability Estimation in IRT Models

After estimating the item parameters using a particular IRT model, the next step is to obtain estimates of the latent trait for each examinee who has responded to the N items. Given a vector of responses to the N items for a person with a latent trait level of θ_j, we can calculate the probability of a correct response, $P(\theta_j, a_i, b_i, c_i)$, and an incorrect response, $Q(\theta_j, a_i, b_i, c_i) = 1 - P(\theta_j, a_i, b_i, c_i)$, to each item given the parameters in our IRT model (in this case the 3PL). When we multiply the probabilities of the items in the response pattern, we obtain the joint likelihood function for the N items as follows:

$$L(\theta_j) = \prod_{i=1}^{N} P_i(\theta_j, a_i, b_i, c_i)^{x_i} Q_i(\theta_j, a_i, b_i, c_i)^{1-x_i}, \tag{5.9}$$

where x_i is the score for person j on item i (either a 0 or a 1). There is a single likelihood function for each unique response pattern. Below we describe three methods widely used for the estimation of latent trait through the joint likelihood function:

1. **Maximum Likelihood Estimation (MLE):** This method aims to find the latent trait that is most likely to given the observed response pattern of a person and the item parameters. That is, the MLE finds the estimate of θ_j that maximizes the likelihood function.

2. **Maximum a Posteriori (MAP):** A common variant of the MLE approach is the Bayesian modal estimation procedure, also called maximum a posteriori, or MAP, where the joint likelihood function is multiplied by an additional curve that represents an assumed population distribution. MAP computes the mode of this posterior distribution as the final estimate of the latent trait.

3. **Expected a Posteriori (EAP):** A variant of the MAP approach is the expected a posteriori, or EAP, which uses the mean instead of the mode from the posterior distribution.

The primary advantage of the Bayesian methods is that they are capable of estimating the latent trait when an examinee has only correct or incorrect responses, whereas the MLE approach cannot provide an estimate of the latent trait when there is zero variability in a response vector. Using the Bayesian methods introduces bias into the estimated parameters through the assumed population distribution. The **mirt** package is capable of estimating latent traits using the MLE, MAP, and EAP approaches via the `fscores` function. In addition, the `fscores` function can employ EAP for sum-scores (Thissen, Pommerich, Billeaud, & Williams, 1995) and weighted likelihood estimation (Warm, 1989). Because the MLE, MAP, and EAP are commonly used by researchers and practitioners, we focus only on these methods below.

In the following example, we use the SAPA data set and the item parameters from the 2PL model that was fit earlier in the chapter to estimate the latent traits using the MLE, MAP, and EAP approaches.

```
latent_mle <- fscores(twopl_fit, method = "ML",
                      full.scores = TRUE, full.scores.SE = TRUE)
latent_map <- fscores(twopl_fit, method = "MAP",
                      full.scores = TRUE, full.scores.SE = TRUE)
latent_eap <- fscores(twopl_fit, method = "EAP",
                      full.scores = TRUE, full.scores.SE = TRUE)
```

To view the MLE latent trait estimates for the first six examinees, we can use the head function with latent_mle.

```
head(latent_mle)
```

```
             F1        SE_F1
[1,] -1.7233250  0.5971554
[2,] -0.7311886  0.4026671
[3,] -0.6771320  0.3986155
[4,] -1.3901749  0.5032752
[5,] -0.7127487  0.4012232
[6,]  1.8175250  0.5662481
```

In the output, the first column (F1) shows the estimated latent trait and the second column (i.e., SE_F1) shows the standard errors. To compare the three estimation methods, we extract the estimated latent traits, which are located in the first column, from the three estimation methods, and combine them in a single data frame named latent.

```
latent <- data.frame(MLE = latent_mle[ ,1],
                     MAP = latent_map[ ,1],
                     EAP = latent_eap[ ,1])
head(latent)
```

```
        MLE         MAP         EAP
1 -1.7233250 -1.3365328 -1.4067370
2 -0.7311886 -0.6307885 -0.6505215
3 -0.6771320 -0.5854772 -0.6028731
4 -1.3901749 -1.1338101 -1.1870565
5 -0.7127487 -0.6153768 -0.6343019
6  1.8175250  1.4356287  1.4935857
```

We see that the three estimation techniques give similar results, with the estimates from MAP and EAP being nearly identical. As we noted earlier, the MLE method cannot estimate latent traits when all responses are either correct or incorrect. In the SAPA data set, there are several examinees who answered all of the items either correctly (e.g., examinees 73, 89, and 103) or incorrectly (e.g., examinee 105), and thus MLE could not estimate the latent trait for these examinees. We show this below by printing these examinees' estimates from the latent data frame.

```
latent[c(73, 89, 103, 105), ]
```

```
      MLE       MAP        EAP
73    Inf  1.985339   2.096258
89    Inf  1.985339   2.096258
103   Inf  1.985339   2.096258
105  -Inf -1.864351  -1.980879
```

To calculate descriptive statistics and correlations among the latent trait estimates, we first remove the examinees with either Inf or -Inf scores in the MLE column using the `is.finite` function and save the rest of the data as `latent_est`. The resulting data frame has 1461 examinees with valid latent trait estimates from the MLE, MAP, and EAP methods.

```
latent_est <- latent[is.finite(latent$MLE), ]
```

Next, we use the `summary` and `apply` functions to obtain descriptive statistics. As a refresher, the `apply` function requires that we specify the data, the margin of the data we would like the function to be applied to (1 for rows and 2 for the columns), and the function we would like to apply. Below we use the `apply` function to calculate the summary statistics and standard deviation (via the `sd` function) of the estimates and specify 2 because they are arranged as columns in our data.

```
apply(latent_est, 2, summary)
apply(latent_est, 2, sd)
```

	MLE	MAP	EAP
Min.	-2.439114424	-1.62922178	-1.725578546
1st Qu.	-0.689406667	-0.59580053	-0.613718783
Median	-0.025201555	-0.02195921	-0.019611185
Mean	0.002659149	0.00357282	0.001146392
3rd Qu.	0.664415606	0.57240513	0.584675590
Max.	2.590186515	1.76241571	1.851380587

```
      MLE       MAP       EAP
1.0112851 0.8021994 0.8345610
```

The mean and median of the latent traits from the three estimation methods are quite similar. However, the range and standard deviations of the latent trait estimates from the MLE approach are greater than those of the other two methods. This is an expected outcome because the MAP and EAP methods are known to shrink the latent trait distribution toward the population mean defined as the prior mean in the estimation process (Reise & Revicki, 2014).

Next, we use the `cor` function to calculate the correlations among the latent trait estimates.

```
cor(latent_est)
```

	MLE	MAP	EAP
MLE	1.0000000	0.9973045	0.9978970
MAP	0.9973045	1.0000000	0.9999447
EAP	0.9978970	0.9999447	1.0000000

The extremely high correlations among the latent trait estimates indicate high concordance between the three estimation methods. However, this may not always be the case as the inclusion of a strong prior distribution is likely

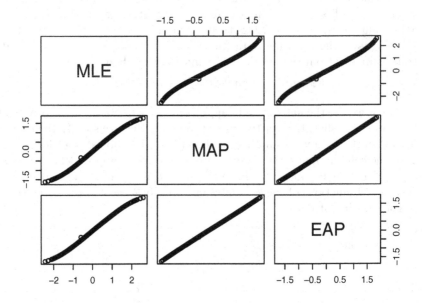

FIGURE 5.13
Scatterplot matrix of latent trait estimates.

to affect the posterior estimates when the MAP and EAP methods are used. To understand the association between the three estimation methods better, we use the `pair` function and create a scatterplot matrix of the latent traits estimated for each method.

```
pairs(latent_est)
```

Figure 5.13 shows that the latent trait estimates from the MAP and EAP methods are highly correlated and linearly related, whereas the latent trait estimates from the MLE method show a slightly nonlinear relationship with the MAP and EAP methods. The nonlinearity is especially noticeable in the tails of their distributions.

Finally, we compute the root mean squared deviation (RMSD), which is the sample standard deviation of the differences between latent trait values estimated by the MLE, MAP, and EAP methods. RMSD can be computed as follows:

$$RMSD = \sqrt{\frac{\sum_{j=1}^{N}(\hat{\theta}_{1j} - \hat{\theta}_{2j})^2}{N}}, \tag{5.10}$$

where $\hat{\theta}_{1j}$ and $\hat{\theta}_{2j}$ are the latent trait estimates for person j from the first and second estimation methods. Note that the order of estimation methods (e.g., MLE-MAP vs. MAP-MLE) does not influence the RMSD value since this difference is squared in the equation. We can calculate RMSD using the rmsd function in **hemp**.

```
rmsd(latent_est$MLE, latent_est$MAP)
```

```
[1] 0.03492328
```

```
rmsd(latent_est$MLE, latent_est$EAP)
```

```
[1] 0.05782216
```

```
rmsd(latent_est$MAP, latent_est$EAP)
```

```
[1] 0.09274544
```

The output indicates that the difference between the latent trait estimates from the MLE and MAP methods is the smallest, whereas the difference between the latent trait estimates from the MAP and EAP methods is the largest.

5.5 Model Diagnostics

We provided a brief introduction to the assumptions of IRT (e.g., unidimensionality, local independence, and invariance of item parameters) at the beginning of this chapter. When these assumptions are met to a reasonable degree, the next step is to select a particular IRT model and apply it to the response data. As in any statistical endeavor, it is possible to observe discrepancies between the selected IRT model and item response data. To provide a basis for the selected IRT model, it is important to perform model diagnostics that can help us identify these potential discrepancies.

In IRT, model diagnostics can be examined at three levels: at the item, person, and model level (i.e., goodness-of-fit of the model). In the following sections, we briefly discuss methods of model diagnostics in IRT and demonstrate the use of these methods using the **mirt** package. For a detailed discussion of model diagnostics in IRT, we encourage readers to review Maydeu-Olivares (2013), Swaminathan, Hambleton, and Rogers (2006), and other resources.

5.5.1   Item Fit

There are generally two approaches for assessing item fit in IRT: item fit statistics and graphical analysis of item fit. The most frequently used item fit statistics include the signed χ^2 test (Orlando & Thissen, 2000), Bock's χ^2 method (Bock, 1972), Yen's $Q1$ statistic (Yen, 1981), the $G2$ statistic (McKinley & Mills, 1985), and the infit and outfit statistics specifically for IRT models from the Rasch family. Some of these methods, such as Bock's χ^2 method, Yen's $Q1$ statistic, and the $G2$ statistic, depend on the comparison of empirical probabilities with model-predicted probabilities. For these methods, significance tests can be used under the null hypothesis that the selected IRT model fits the data adequately. In its general form, a χ^2 item fit statistic for a dichotomous item can be computed as

$$\chi^2 = \sum_{j=1}^{J} N_j \frac{(O_j - E_j)^2}{E_j(1 - E_j)}, \tag{5.11}$$

where j indexes the latent trait interval (e.g., $\theta = -2$ to $\theta = -1.5$), O_j is the observed (i.e., empirical) probability of a correct response in latent trait interval j, E_j is the expected (i.e., theoretical) probability of a correct response in latent trait interval j based on the fitted IRT model, and N_j is the number of examinees located in latent trait interval j.

In Bock's χ^2 method, each latent trait interval has equal size and the median latent trait in each interval is used for computing the expected proportion correct. In contrast, Yen's $Q1$ statistic requires 10 latent trait intervals with equal frequency and the expected probability becomes the mean of the probabilities of a correct response. Both χ^2 statistics have degrees of freedom equal to $J - m$, where m is the number of item parameters in the IRT model (e.g., $m = 2$ for the 2PL model).

Orlando and Thissen (2000) argued that the presence of significant item misfit may lead to biased latent trait estimates, thereby affecting the test of item fit in the χ^2 methods. Therefore, Orlando and Thissen (2000) suggested that examinees be grouped according to raw scores instead of latent trait estimates. This alternative method is called the signed χ^2 statistic, which is also known as the $S - X^2$ statistic. The signed χ^2 statistic for a given item can be written as

$$S - X^2 = \sum_{k}^{n-1} N_k \frac{(O_k - E_k)^2}{E_k(1 - E_k)}, \tag{5.12}$$

where n is the number of items on the test, k is the raw score (i.e., the number of correct responses), and the other elements are the same as those in Equation

5.11. The degrees of freedom for the signed χ^2 statistic is $(n-1) - m$, where m is again the number of item parameters in the IRT model.

Unlike the χ^2 item fit statistics, the $G2$ statistic is based on a likelihood ratio test. Using the same elements in Equation 5.11, the $G2$ statistic can be computed as follows:

$$G^2 = 2 \sum_{j=1}^{J} N_j \left(O_j log\frac{O_j}{E_j} + (1 - O_j)log\frac{1 - O_j}{1 - E_j} \right). \tag{5.13}$$

The $G2$ statistic has degrees of freedom equal to $J - m$. As mentioned earlier, Bock's χ^2 statistic, Yen's $Q1$ statistic, and the $G2$ statistic work under the null hypothesis that the item shows adequate fit with the IRT model. Therefore, rejecting the null hypothesis for a particular item suggests that the item may not fit the model adequately.

In the following example, we use the `itemfit` function from the **mirt** package to examine the item fit statistics for the item parameters from the Rasch model. In the `itemfit` function, we specify `fit_stats = c("S_X2", "G2")` to obtain the $S - X^2$ and $G2$ item fit statistics. The `impute` option specifies the number of imputations to perform when missing data are present. We use `impute = 10` to create ten imputed data sets because there are several missing responses in the **SAPA** data set. We save the results as `rasch_itemfit` and print the first six rows of the results with the `head` function.

```
rasch_itemfit <- itemfit(rasch_fit,
                  fit_stats = c("S_X2", "G2"),
                  impute = 10)
head(rasch_itemfit)
```

	item	G2	df.G2	p.G2	S_X2	df.S_X2	p.S_X2
1	reason.4	10.888	9	0.344	20.597	12	0.057
2	reason.16	9.007	9	0.443	10.802	12	0.546
3	reason.17	15.409	9	0.131	19.920	12	0.069
4	reason.19	14.018	9	0.166	26.702	12	0.009
5	letter.7	5.458	9	0.788	12.490	12	0.407
6	letter.33	14.195	9	0.163	27.096	13	0.012

In the output, the columns G2, df.G2, and p.G2 indicate the $G2$ statistic, the degrees of freedom for the $G2$ statistic, and the p-value for the $G2$ statistic. Similarly, the columns S_X2, df.S_X2, and p.S_X2 indicate the $S - X^2$ statistic, the degrees of freedom for the $S - X^2$ statistic, and the p-value for the $S - X^2$ statistic. If we use an alpha level of $\alpha = .05$ as the significance level, then p-values below .05 for a given item would suggest that the item does not fit the Rasch model.

The `itemfit` function also provides other item fit measures such as the Z_h, infit, and outfit statistics, which quantify the magnitude of overall discrepancy between the selected IRT model and the response data. The infit and outfit statistics are only available for the Rasch family of models (e.g., Rasch model, partial credit model, and rating scale model). In the following example, we show how to obtain the Z_h, infit, and outfit statistics for the Rasch model, using `fit_stats = c("Zh", "infit")` (the output is not printed here).

```
itemfit(rasch_fit,
        fit_stats = c("Zh", "infit"),
        impute = 10)
```

In the literature, there are different recommendations regarding the cutoff values for interpreting the Z_h, infit, and outfit statistics. We will not discuss the details of these recommendations in this chapter. In general, we do not recommend that decisions about items be made based solely on the item fit statistics. These statistics often disagree (as can be seen above for the reason.19 item) and should be used with caution. Therefore, these statistics presented above are shown just to make our readers aware of them and for didactic purposes only.

The `itemfit` function can also be used for creating an empirical plot for the items, which can be more useful than the item fit statistics in detecting item misfit. In the following example, we create an empirical plot for item 1 in the `SAPA` data set. The plot shows the empirical data points and the theoretical ICC for the Rasch model together. The closer the empirical data follow the ICC, the better item fit, which appears to be the case for item 1 based on Figure 5.14.

```
itemfit(rasch_fit, empirical.plot = 1)
```

5.5.2   Person Fit

In the context of IRT, person fit refers to the alignment between an examinee's response pattern and the IRT model selected for modeling the response data. Person-fit indices are used to assess the validity of the selected model at the examinee level and the meaningfulness of the estimated latent trait levels (Embretson & Reise, 2000). In the literature, there are various person-fit indices proposed for the IRT models. In this section, we focus, specifically, on the standardized fit index called Z_h, originally proposed by Drasgow, Levine, and Williams (1985). The Z_h statistic for examinee j with latent trait level θ_j can be computed as:

$$Z_h = \sum \left[\mathrm{Log}L | \theta_j - \sum E(\mathrm{Log}L | \theta_j) \right] / \sqrt{\left(\sum V(\mathrm{Log}L | \theta_j) \right)}, \qquad (5.14)$$

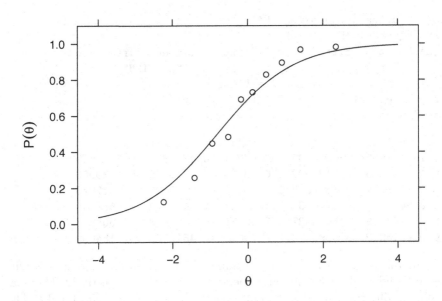

Empirical plot for item 1

FIGURE 5.14
Empirical plot for item 1.

where LogL is the log-likelihood value for the examinee j's response pattern, $E(\text{Log}L|\theta_j)$ is the average log-likelihood value for the sampling distribution of log-likelihoods conditional on θ_j, and $V(\text{Log}L|\theta_j)$ is the variance of the sampling distribution of log-likelihood values. Since Z_h is a standardized statistic that follows a standard normal distribution, the expected value for Z_h is zero when the examinee's response pattern is aligned with the selected IRT model. Large negative Z_h values (e.g., $Z_h < -2$) indicate person misfit. Large positive Z_h values indicate that the likelihood for the examinee's response pattern is higher than the predicted likelihood based on the selected IRT model (Embretson & Reise, 2000), which might also be problematic for latent trait estimation.

In the following example, we use the `personfit` function from the **mirt** package to obtain the Z_h statistic for the examinees' response patterns in the **SAPA** data set. The `personfit` function requires the response data to have no missing data. Therefore, we first remove examinees with missing data from the **SAPA** data set, fit the Rasch model again using the complete data set, calculate the person-fit statistics, and print fit statistics for the first few examinees. This time, we specify the `verbose = FALSE` argument to prevent `mirt` from printing information about the change in the log-likelihoods to the R console.

```
SAPA_nomiss <- na.omit(SAPA)
rasch_mod <- "F = 1 - 16"
rasch_fit <- mirt(data = SAPA_nomiss, model = rasch_mod,
                  itemtype = "Rasch", SE = TRUE,
                  verbose = FALSE)
rasch_personfit <- personfit(rasch_fit)
head(rasch_personfit)
```

```
    outfit    z.outfit      infit    z.infit         Zh
1 1.2568668  0.5629182  1.0167307  0.1759854 -0.2366559
2 1.1913379  0.5135602  1.0413354  0.2503205 -0.3039903
3 0.7145792 -0.5621505  0.8733401 -0.5890823  0.6562943
4 0.5611105 -0.3596991  0.7938596 -0.4143222  0.5659395
5 0.7357619 -0.5034437  0.8989875 -0.4512893  0.5565336
6 0.4846912 -0.6398484  0.7733820 -0.4544976  0.6154669
```

In the output, the first four columns, outfit, z.outfit, infit, and z.infit, correspond to the infit and outfit person-fit statistics, respectively. The final column, Zh, can be used to examine person fit based on the Z_h statistic. The results show that the first six examinees' response patterns are aligned with the Rasch model because their Z_h statistics are larger than -2. We can also easily create a histogram of the distribution of the Z_h statistics using the `hist` function. We use `abline(v = -2, lwd = 2, lty = 2)` to draw a vertical, dashed line at $Z_h = -2$ as a reference point for identifying the examinees with person misfit. Figure 5.15 shows that the Z_h statistic is larger than -2 for most examinees, suggesting most examinees fit the selected IRT model.

```
hist(rasch_personfit$Zh, xlab="Zh Statistic")
abline(v = -2, lwd = 2, lty = 2)
```

Finally, we subset and print the examinees with Z_h statistics less than -2 using the `rownames` function. We see that there were 40 examinees that did not fit the Rasch model using the `nrow` function.

```
rasch_misfits <- subset(rasch_personfit, Zh < -2)
rownames(rasch_misfits)
```

```
 [1] "41"   "63"   "65"   "98"   "214"  "227"  "268"  "269"
 [9] "275"  "354"  "368"  "393"  "411"  "445"  "466"  "470"
[17] "503"  "524"  "534"  "537"  "560"  "566"  "597"  "619"
[25] "668"  "684"  "735"  "804"  "806"  "818"  "849"  "889"
[33] "946"  "1017" "1092" "1109" "1326" "1337" "1394" "1464"
```

```
nrow(rasch_misfits)
```

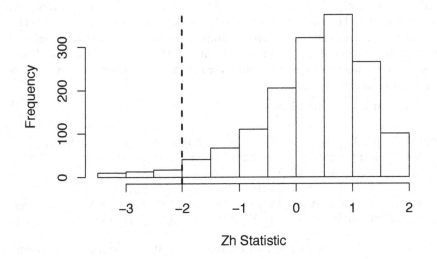

FIGURE 5.15
Distribution of the Zh statistic for the Rasch model.

[1] 40

5.5.3 Model Selection

Model selection in IRT depends on both theory underlying the assessment and the characteristics of the response data. For example, assuming that all items have the same level of discrimination, one can prefer to use the 1PL model (or Rasch model) over the other models (e.g., 2PL and 3PL). However, if the discrimination parameters of the items vary significantly, constraining item discrimination parameters to be equal across all items may not be the best decision. Therefore, one can fit different IRT models to the same data and compare the fit of these models. In unidimensional IRT models, this comparison can be done in several ways. Because the 1PL, 2PL, and 3PL IRT models are considered nested (e.g., 1PL is nested within the 2PL if we constrain the discrimination parameters to be equal across the items and within the 3PL if we add the additional constraint that the pseudo-guessing parameters are zero across the items), then likelihood-based statistics can be used for comparing the overall fit of the models. Model comparison in the **mirt** package can be done with the **anova** function.

In the following example, we use the previously fitted models for the **SAPA** data set to show how to perform model comparisons. The first comparison is between the 1PL model and the 2PL model. We want to know whether the 2PL model provides a significant improvement over the 1PL model with regards to the model fit. We previously fit the 1PL and the 2PL models and saved the results as **onepl_fit** and **twopl_fit**. To compare the two models, we use the **anova** function.

```
anova(onepl_fit, twopl_fit)
```

```
Model 1: mirt(data = SAPA, model = onepl_mod, SE = TRUE)
Model 2: mirt(data = SAPA, model = twopl_mod, itemtype = "2PL",
              SE = TRUE)
        AIC      AICc     SABIC      BIC    logLik      X2  df    p
1 26566.50 26566.91 26603.10 26657.11 -13266.25     NaN NaN  NaN
2 26464.79 26466.21 26533.69 26635.34 -13200.40  131.71  15    0
```

Several different model fit indices, the AIC, AICc (a sample size corrected AIC), SABIC, and BIC, are presented in the output. The smaller these fit indices, the better the model fits the data. In the output, we see that all of the fit indices are smaller for the 2PL model, suggesting that the 2PL model fits the **SAPA** data set better than the 1PL model. The **anova** function also performs a likelihood ratio test based on the log likelihood values from the two models. Two times the absolute difference in log-likelihood values of the two models has a chi-square (χ^2) distribution, with degrees of freedom equal to the difference in the number of estimated parameters between the two models, under the null hypothesis that there is no difference in model fit between the two models. In other words, the model with more parameters is assumed to fit the data no better than the model with fewer parameters. If the p-value for the χ^2 statistics is less than .05 (using the de facto $\alpha = .05$), then we can conclude that the more complex model fits better than the simpler model.

In the output, we see that $\chi^2_{15} = 131.71, p < .001$, which suggests that the 2PL model provided better fit than the 1PL model to the **SAPA** data set. This finding is consistent with the model fit indices that favor the 2PL model over the 1PL model. In the following step, we compare the fit of the 2PL and the 3PL models. This comparison tests whether including the pseudo-guessing parameter in the model provides an improvement in model fit. All of the model fit indices are smaller for the 3PL model, which favors the 3PL model over the 2PL model. The likelihood ratio test $(\chi^2_{16} = 160.058, p < .001)$ is statistically significant, which means that the 3PL model provided a statistically significant improvement in model fit over the 2PL model.

```
anova(twopl_fit, threepl_fit)
```

```
Model 1: mirt(data = SAPA, model = twopl_mod, itemtype = "2PL",
              SE = TRUE)
Model 2: mirt(data = SAPA, model = threepl_mod, itemtype = "3PL",
              SE = TRUE)
       AIC      AICc    SABIC      BIC    logLik     X2  df    p
1 26464.79 26466.21 26533.69 26635.34 -13200.40    NaN NaN  NaN
2 26336.74 26339.92 26440.08 26592.56 -13120.37 160.058 16    0
```

Our final comparison is between the 3PL model and the 4PL model. The results again suggest that the more complex model (i.e., the 4PL model) fits the SAPA data set better than the less complex model.

```
anova(threepl_fit, fourpl_fit)
```

```
Model 1: mirt(data = SAPA, model = threepl_mod, itemtype = "3PL",
              SE = TRUE)
Model 2: mirt(data = SAPA, model = fourpl_mod, itemtype = "4PL",
              SE = TRUE)
       AIC      AICc    SABIC      BIC    logLik     X2  df    p
1 26336.74 26339.92 26440.08 26592.56 -13120.37    NaN NaN  NaN
2 26259.96 26265.66 26397.75 26601.06 -13065.98 108.778 16    0
```

It should be noted that the models with more parameters often tend to fit the data better than the models with fewer parameters. Therefore, the comparison of overall model fit may not always lead to the right decision about which model should be selected. As we explained earlier, model selection should not be based merely on tests of model fit or theoretical assumptions/expectations about the assessment. We encourage our readers to review item parameter estimation and model fit as part of the test design, instead of a standalone step to determine the destiny of the assessment.

5.6 Summary

In this chapter, we introduced the unidimensional IRT models for dichotomous item responses and demonstrated how to estimate these models using the mirt function from the **mirt** package. The IRT models presented in this chapter are suitable for dichotomously scored items (e.g., right or wrong, agree or disagree, yes or no) in a unidimensional test structure. The mirt function is capable of estimating common IRT models for dichotomous items. Additionally, it allows users to create more specialized or constrained models by adding additional elements (e.g., constraining the lower asymptote parameter to be the same across all items). In addition to the `mirt` function, we illustrated several graphical tools from the **mirt** package to create IRT plots at the item and test levels. Then we showed how to obtain latent trait estimates using the `fscores` function in the **mirt** package. This function is capable of estimating latent trait levels using the MLE, EAP, and MAP approaches. Finally, we provided a section on model diagnostics in IRT. This section demonstrates the `itemfit`, `personfit`, and `anova` functions to examine item fit, person fit, and overall model fit. We encourage interested readers to checkout the "Psychometrics Task View" (Mair & Hatzinger, 2007b) on **CRAN** for a description of other R packages available for IRT modeling (see Chapter 12 for more information about CRAN Task Views). In the following chapter, we focus on the unidimensional IRT models for polytomously scored items.

6

Item Response Theory for Polytomous Items

6.1 Chapter Overview

In Chapter 5, we introduced the unidimensional IRT models for dichotomously scored items with two response categories (e.g., right or wrong, yes or no, agree or disagree). When items consist of more than two response categories (either ordinal or nominal categories), these items are referred to as polytomously scored items (or polytomous items). In Chapter 6, we focus on the item response models for polytomously scored items. The IRT models for polytomous items can be seen as generalized forms of the IRT models for dichotomously scored items (e.g., Rasch model, 1PL model, or 2PL model). Similar to Chapter 5, Chapter 6 presents various IRT models designed for polytomously scored items and demonstrates how to estimate these models in R. Polytomous IRT models covered in this chapter include the partial credit model (Masters, 1982), the generalized partial credit model (Muraki, 1992), the rating scale model (Andrich, 1978), the graded response model (Samejima, 1969), the nominal response model (Bock, 1972), and the nested logit model (Suh & Bolt, 2010). We will use the **mirt** package (Chalmers, 2012) to demonstrate IRT model estimation for polytomously scored items.

In this chapter, we categorize the polytomous IRT models into three groups based on their scale of measurement (i.e., ordinal or nominal) and whether they are in the Rasch family of models:

1. Polytomous Rasch models for ordinal items

2. Polytomous non-Rasch models for ordinal items

3. Polytomous IRT models for nominal items

Other researchers have also proposed similar categorizations for polytomous IRT models (e.g., De Ayala, 2013).

6.2 Polytomous Rasch Models for Ordinal Items

The partial credit model (Masters, 1982) and the rating scale model (Andrich, 1978) are considered polytomous forms of the Rasch model. These models assume that the adjacent categories in a polytomously scored ordinal item (e.g., strongly disagree, disagree, agree, or strongly agree) are in fact two dichotomous categories, as in the dichotomous IRT models. Therefore, these models use the same logic of dichotomous IRT models to estimate the probability of obtaining the higher category (e.g., strongly agree over agree) based on the respondent's latent trait level. Despite using the same mechanism to estimate item parameters, the partial credit model (Masters, 1982) and the rating scale model (Andrich, 1978) differ from each other with regard to the estimation of category boundaries (i.e., thresholds). The following sections will briefly describe these models and demonstrate how to estimate them using the **mirt** package (Chalmers, 2012) in R.

6.2.1 Partial Credit Model

The partial credit model (PCM; Masters, 1982), which is also known as the adjacent category logit model, is a polytomous form of the Rasch model. PCM assumes that the polytomous items consist of multiple ordered categories. The model focuses on adjacent categories when estimating the thresholds (i.e., difficulty) between the ordered response categories. Given that an item has K ordered response categories, PCM estimates $K-1$ thresholds for the item. For example, if an item has four response categories (strongly disagree, disagree, agree, or strongly agree), PCM would estimate three thresholds ($\delta_1, \delta_2, \delta_3$) between the adjacent categories. Figure 6.1 shows the thresholds for a four-category item.

The probability of obtaining X_i points ($X_i = 0, 1, \ldots, m_i$) on item i for PCM can be written as

$$P(X_i|\theta, \delta_{ih}) = \frac{exp\left[\sum_{h=0}^{X_i}(\theta - \delta_{ih})\right]}{\sum_{k=0}^{m_i} exp\left[\sum_{h=0}^{k}(\theta - \delta_i h)\right]}, \tag{6.1}$$

where θ is the latent trait, δ_{ih} is the step parameter (also known as step difficulty) that represents the relative difficulty in obtaining h points over $(h-1)$ points (De Ayala, 2013). Because the subscript m_i sets the maximum response category for each item individually, PCM allows the number of response categories to differ across items.

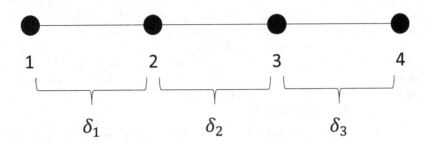

FIGURE 6.1
Thresholds between the four ordered response categories.

It should be noted that PCM does not require the thresholds to follow the same order as the response categories. Because PCM considers adjacent categories in each step, the adjacent response categories are treated as a series of dichotomous items, but without order constraints beyond adjacent categories. Continuing from the same example in Figure 6.1, the threshold between response categories 1 and 2 (i.e., δ_1) can be larger than the threshold (δ_2) between response categories 2 and 3.

This feature makes PCM very flexible when estimating item parameters for Likert-type rating scales and surveys because sometimes respondents may choose higher categories more often than lower categories. For example, assume a survey question asks how often a person uses his or her smartphone during the day, and the response options are never, very rarely, sometimes, and very often. If most respondents tend to use their phones very frequently during the day, selecting the response options of sometimes and very often are more likely to be selected than either never or very rarely. Thus, the threshold between sometimes and very often can be smaller than the threshold between never and very rarely.

Unlike Likert-type items, polytomous items that measure achievement or aptitude are expected to follow the same order as the response categories. For example, in a math problem with possible scores of 0, 1, and 2, obtaining 2 points would be more difficult than obtaining 0 or 1 point. Therefore, we would expect the threshold between 1 and 2 to be larger than the threshold between 0 and 1. Depending on item content and the latent trait being measured, users should make sure that the order of response categories from PCM is functioning as expected.

In the following example, we use the **rse** data set from the **hemp** package to demonstrate how to fit a PCM. The **rse** data set comes from the Rosenberg self-esteem scale (Rosenberg, 1965) that measures individuals' self-esteem. The scale consists of 10 statements with four response options: 1 = strongly disagree, 2 = disagree, 3 = agree, and 4 = strongly agree. The `rse` data set consists of a random sample of 1000 respondents who completed all of the items on the instrument. Because the statements in items 3, 5, 8, 9, and 10 are negatively worded (e.g., Q3: All in all, I am inclined to feel that I am a failure), responses to these items have been reverse-coded (i.e., 1 = strongly agree, 2 = agree, 3 = disagree, 4 = strongly disagree). Thus, higher scores for all items indicate greater self-esteem. More details about the `rse` data set can be found in the **hemp** package.

Before we begin the analysis, we first activate the **hemp** and **mirt** packages (Chalmers, 2012) using the `library` function.

```
library("hemp")
library("mirt")
```

As we have done for the dichotomous IRT models in Chapter 5, we begin by defining the latent trait (called `selfesteem` in the example) using the first 10 variables in the `rse` data set (`rse[, 1:10]`), which correspond to the items in the Rosenberg self-esteem scale. Next, we define and save the model as `pcm_mod` and estimate it using the `mirt` function. Although we set `itemtype = "Rasch"`, the `mirt` function recognizes that the items may have more than two response categories, and thus fits PCM. In the `rse` data set, all of the items have four response categories. However, as mentioned earlier, the number of response categories may vary across items when PCM is used. Once the model is estimated, we extract the results using the `coef` function and save them as `pcm_params`.

```
pcm_mod <- "selfesteem = 1 - 10"
pcm_fit <- mirt(data = rse[, 1:10], model = pcm_mod,
                itemtype = "Rasch", SE = TRUE)
pcm_params <- coef(pcm_fit, IRTpars = TRUE, simplify = TRUE)
```

As we mentioned in Chapter 5, the `coef` function saves the item parameters and additional information as a list. Thus, we select the item parameters from this list, save them as a new data frame called `pcm_items`, and then print the item parameters.

```
pcm_items <- as.data.frame(pcm_params$items)
pcm_items
```

	a	b1	b2	b3
Q1	1	-2.861206	-1.64028435	0.9910196
Q2	1	-3.053835	-2.16455743	1.0889792
Q3	1	-2.221324	-0.53599085	1.7627099
Q4	1	-3.349555	-1.45422358	1.8059802
Q5	1	-1.848675	-0.07989451	1.4722142
Q6	1	-2.182504	-0.16515281	2.0232217
Q7	1	-1.782826	0.13784822	2.3143204

```
Q8  1 -1.581019  0.87474864 1.9708572
Q9  1 -1.235317  1.21761688 2.0240591
Q10 1 -1.398785  0.61305040 1.1605091
```

In the output, the names of the items in the `rse` data set are printed as the row names. The first column shows the discrimination parameter (a), which is equal to 1 for all items because PCM, similar to the Rasch model, constrains the discrimination parameter to be 1. In the next set of columns (b1 to b3), we see the estimated thresholds (i.e., step) parameters. Because the items in the `rse` data set had four response options, PCM estimated three threshold parameters for each item. The columns labeled as b1 to b3 in the output correspond to the δ parameters in Equation 6.1.

Next, we use the `plot` function in the `mirt` package to examine the items visually. Unlike dichotomous IRT models, polytomous IRT models have option characteristic curves (OCCs), which can be considered as an extension of ICCs for polytomous items. Because the items have more than two response categories, there are multiple OCCs plotted per item. Each curve represents the probability of selecting a particular response option as a function of the latent trait (θ).

```
plot(pcm_fit, type = "trace", which.items = c(2, 5),
     par.settings = simpleTheme(lty = 1:4, lwd = 2),
     auto.key = list(points = FALSE, lines = TRUE, columns = 4))
```

When we call the `plot` function, we provide the fitted model (`pcm_fit`) and specify the type of plot we want to use (`type = "trace"`). To demonstrate how we would plot the OCCs for only two items, we specify `which.items = c(2, 5)` to plot Q2 and Q5 in the `rse` data set. If we removed this argument from the `plot` function, by default it would draw the OCCs for all items in a single, trellis plot. Next, we specify some additional settings for the plot using the `par.settings` argument and the `simpleTheme` function. We set `lty = 1:4` to request different line types for each OCC, and to make the curves thicker, we set `lwd = 2`, which stands for line width. Finally, we create a legend using the `auto.key`. This legend has 4 columns that show the lines and not the points for our OCCs. In Figure 6.2, the response categories are labeled as P1 to P4. For both items, the OCCs follow the same expected order as the response categories. The OCC for the first response option (P1) is on the far left side of the plot, whereas the last response option (P4) is located on the far right side of the plot. The other OCCs in the middle are also ordered properly.

Next, we plot the IIF to show the amount of information that each item explains as a function of their latent trait level. We use the `plot` function again but this time we request the IIFs by setting `type = "infotrace"`. In the `par.settings` option, we again change the width of the lines to make

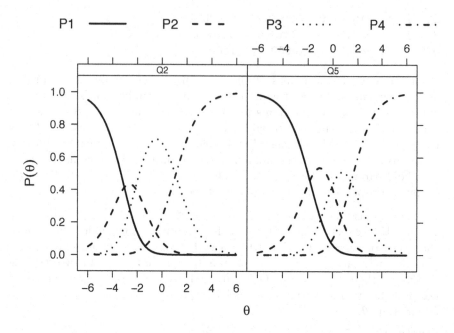

FIGURE 6.2
Option characteristic curves for items 2 and 5 for PCM.

them more visible. Figure 6.3 shows the IIF plot for items 2 and 5 in the `rse` data set.

```
plot(pcm_fit, type = "infotrace", which.items = c(2, 5),
     par.settings = simpleTheme(lwd = 2))
```

In addition to IIFs for the individual items, we can also plot the total amount of information (i.e., test information function — TIF) available from the items and the distribution of the conditional standard error of measurement (cSEM) in the `rse` data set. In the following example, we show how to plot the TIF and cSEM over a latent trait range of -6 to 6. We use `type = "info"` for the TIF plot and `type = "SE "` for the cSEM plot.

```
plot(pcm_fit, type = "info", theta_lim = c(-6, 6))
plot(pcm_fit, type = "SE", theta_lim = c(-6, 6))
```

6.2.2   Rating Scale Model

The rating scale model (RSM; Andrich, 1978) can be viewed as a restricted form of PCM. RSM works well with instruments where all the items have

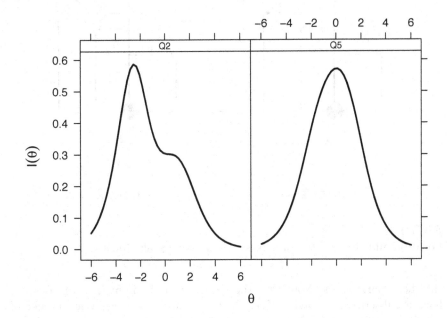

FIGURE 6.3
Item information functions for items 2 and 5 for PCM.

the same structural response form (i.e., the response format is assumed to function in the same way across all the items). A common example of this are instruments that contain Likert scales. RSM assumes that there is a series of ordinal thresholds that separate the adjacent ordinal response categories from each other (De Ayala, 2013). Each item gets a unique location parameter, but the differences between the response categories around the location parameter are constrained to be equal across all items. Therefore, the items in the instrument differ in terms of their overall locations while the spread of category thresholds within the items remains the same.

For the RSM, the probability of selecting category c ($c = 0, 1, \ldots, m$) for item i may be written as:

$$
P(X_{ic}|\theta, \lambda_i, \delta_1, \ldots, \delta_m) = \frac{exp\left[\sum_{j=0}^{c}\left(\theta - (\lambda_i + \delta_j)\right)\right]}{\sum_{h=0}^{m} exp\left[\sum_{j=0}^{h}\left(\theta - (\lambda_i + \delta_j)\right)\right]}, \tag{6.2}
$$

where λ_i is the location parameter for item i and $\delta_1, \ldots, \delta_m$ are the category

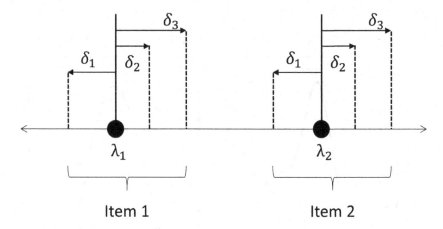

Item 1   Item 2

FIGURE 6.4
Location and threshold parameters of two rating scale items.

threshold parameters. Note that the category of the item (i.e., c) does not have a subscript i, implying that all items must have the same number of categories, unlike the PCM. Figure 6.4 demonstrates two rating scale items on the latent trait continuum. Although item 1 and item 2 differ with regard to their overall location (λ), the spread of the category thresholds (δ) are equivalent across the two items.

In the following example, we fit the RSM to the **rse** data set. All of the items in the **rse** data set have four categories, which allows us to estimate the same number of category threshold parameters for all the items. This time when we define our model, we specify itemtype = "rsm" to estimate the RSM. The itemtype = "rsm" option constrains the discrimination parameters and the category thresholds to be the same across all items. We save the estimation results for RSM as **rsm_fit**, extract the results using the **coef** function, then save the estimated item parameters as **rsm_items**, and print them.

```
rsm_mod <- "selfesteem = 1 - 10"
rsm_fit <- mirt(data = rse[,1:10], model = rsm_mod,
                itemtype = "rsm")
rsm_params <- coef(rsm_fit, simplify = TRUE)
rsm_items <- as.data.frame(rsm_params$items)
rsm_items
```

	a1	b1	b2	b3	c
Q1	1	-3.099763	-1.3378	0.7764022	0.0000000
Q2	1	-3.099763	-1.3378	0.7764022	0.1821878
Q3	1	-3.099763	-1.3378	0.7764022	-0.8676205

```
Q4   1 -3.099763 -1.3378 0.7764022 -0.2935097
Q5   1 -3.099763 -1.3378 0.7764022 -1.0872174
Q6   1 -3.099763 -1.3378 0.7764022 -1.1154409
Q7   1 -3.099763 -1.3378 0.7764022 -1.4301361
Q8   1 -3.099763 -1.3378 0.7764022 -1.7372333
Q9   1 -3.099763 -1.3378 0.7764022 -2.0159556
Q10  1 -3.099763 -1.3378 0.7764022 -1.4512959
```

In the output, a1 is the item discrimination parameter, which is fixed to 1 for all items because RSM, as a polytomous form of the Rasch model, requires that all items have the same discrimination parameter. The following columns (i.e., d1 through d3) represent the category threshold parameters, which are the same for all of the items, and c is the location parameter estimated uniquely for each item in the rse data set. The estimated location parameters show that the easiest item was Q9 and the hardest item was Q2. To see a visual comparison of these two items, we plot the OCCs for items 2 and 9. The code for creating this plot is the same as before for PCM, except that we specify which.items = c(2, 9) instead.

```
plot(rsm_fit, type = "trace", which.items = c(2, 9),
     par.settings = simpleTheme(lty = 1:4, lwd = 2),
     auto.key = list(points = FALSE, lines = TRUE, columns = 4))
```

Figure 6.5 shows that choosing higher response categories (e.g., 4 = strongly agree or 3 = agree) for Q9 is easier than selecting the same response options for Q2. We can also draw a plot of TIF and cSEM for RSM specifying type = "info" for the TIF plot and type = "SE" for the cSEM plot.

```
plot(rsm_fit, type = "info", theta_lim = c(-6, 6))
plot(rsm_fit, type = "SE", theta_lim = c(-6, 6))
```

6.3 Polytomous Non-Rasch Models for Ordinal Items

The generalized partial credit model (Muraki, 1992) and the graded response model (Samejima, 1969) can be viewed as polytomous forms of the two-parameter (2PL) IRT model. Unlike polytomous Rasch models, the generalized partial credit model and the graded response model assume that items vary in terms of their discrimination levels in distinguishing examinees or respondents with low and high latent traits. These two models are slightly different from each other based on the conceptualization of the option characteristic curves (De Ayala, 2013). The following sections will briefly describe

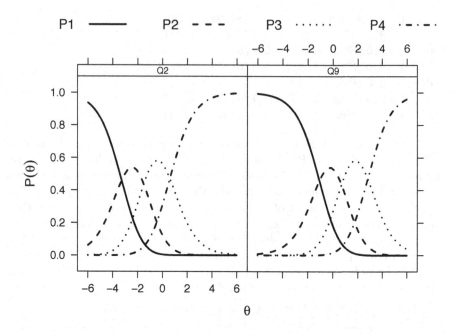

FIGURE 6.5
Option characteristic curves for items 2 and 9 for RSM.

these models and demonstrate how to estimate them using the **mirt** package
(Chalmers, 2012) in R.

6.3.1 Generalized Partial Credit Model

The generalized partial credit model (GPCM; Muraki, 1992) is similar to PCM
regarding how it conceptualizes option characteristic curves, while it is also
similar to the 2PL model in that item discrimination parameters may vary
across the items. Instead of fixing the item discrimination parameter to 1 for
all items, GPCM estimates a unique item discrimination parameter for each
item. According to Muraki (1992), item discrimination parameters indicate
the degree to which categorical responses vary among items as the latent trait
changes. The probability of obtaining X_{ik} points ($X_{ik} = 0, 1, \ldots, m_i$) on item
i for GPCM can be written as

$$P(X_{ik}|\theta, a_i, \delta_{ik}) = \frac{exp\left[\sum_{h=1}^{X_{ik}} a_i(\theta - \delta_{ih})\right]}{\sum_{c=1}^{m_i} exp\left[\sum_{h=1}^{c} a_i(\theta - \delta_i h)\right]}, \quad (6.3)$$

where a_i is the discrimination parameter for item i and the remaining terms are the same as those in Equation 6.1. Similar to PCM, the thresholds (δ_{ik}) are not restricted to be in the same order as the response categories.

In the following example, we estimate the category thresholds as well as the item discrimination parameters for the items in the rse data set using GPCM. Instead of itemtype = "Rasch", this time we use itemtype = "gpcm" to estimate the item parameters based on the GPCM.

```
gpcm_mod <- "selfesteem = 1 - 10"
gpcm_fit <- mirt(data = rse[, 1:10], model = gpcm_mod,
                 itemtype = "gpcm", SE = TRUE)
gpcm_params <- coef(gpcm_fit, IRTpars = TRUE, simplify = TRUE)
gpcm_items <- gpcm_params$items
gpcm_items
```

	a	b1	b2	b3
Q1	1.8238891	-1.7202941	-0.95938961	0.5885070
Q2	1.7307657	-1.8564721	-1.28862013	0.6510270
Q3	1.8710897	-1.3231835	-0.31308686	1.0425363
Q4	1.2811061	-2.2085974	-1.00156932	1.2138474
Q5	1.4939839	-1.1502285	-0.04606415	0.9041157
Q6	2.7132868	-1.2057255	-0.09730049	1.1220286
Q7	2.3609630	-1.0210663	0.07245970	1.3106380
Q8	0.8945663	-1.1800257	0.81206980	1.2679593
Q9	1.4911761	-0.7707405	0.78235286	1.2329194
Q10	1.9005838	-0.8381950	0.33153091	0.7308140

Unlike the estimated item parameters for the PCM, the first column of the output from the GPCM (i.e., item discrimination) is not fixed to 1. Instead, each item has a unique item discrimination parameter. The output shows that the items in the rse data set vary somewhat in terms of item discrimination. The remaining 3 columns show the estimated category threshold parameters from theGPCM.

To demonstrate the impact of the varying item discrimination parameters in the GPCM, we plot the OCCs for Q6 and Q8. Unlike in Figure 6.2, we see that the items estimated from GPCM do not have the same slopes (i.e., discrimination). Figure 6.6 shows that the slopes of the OCCs for Q6 are steeper than the slopes of the OCCs for Q8 as a result of higher item discrimination in Q6.

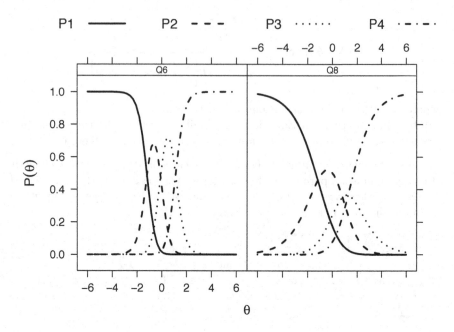

FIGURE 6.6
Option characteristic curves for items 6 and 8 for GCPM.

```
plot(gpcm_fit, type = "trace", which.items = c(6, 8),
     par.settings = simpleTheme(lty = 1:4, lwd = 2),
     auto.key=list(points = FALSE, lines = TRUE, columns = 4))
```

As we have shown for PCM and RSM, we can also draw the plots of TIF and cSEM for GPCM specifying `type = "info"` for the TIF plot and `type = "SE"` for the cSEM plot.

```
plot(gpcm_fit, type = "info", theta_lim = c(-6, 6))
plot(gpcm_fit, type = "SE", theta_lim = c(-6, 6))
```

6.3.2　Graded Response Model

The graded response model (GRM; Samejima, 1969), also known as the cumulative logit model, is a polytomous extension of the 2PL model. The GRM is suitable for items with a clear underlying response continuum. The GRM models the probability of a given response category or higher by following the same order as the response options. In the GRM, each response category contributes some information to a person's probability of selecting a particular response

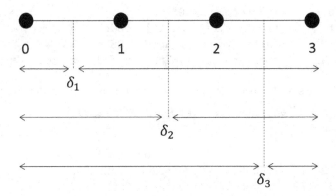

FIGURE 6.7
Cumulative thresholds between the four response categories.

category. For an item with K ordered response categories, the GRM creates $K - 1$ dichotomous items by splitting the response categories in a cumulative fashion. Each of these artificial dichotomous items has a unique difficulty parameter but shares the same discrimination parameter. For example, consider a polytomous item with four ordered response categories ($X = 0, 1, 2, or 3$). Figure 6.7 shows the three thresholds ($\delta_1, \delta_2, \delta_3$) that split the four response categories in a cumulative way. That is, each threshold parameter shows the latent trait level needed to have a 50% chance of selecting a particular response category or higher.

The probability of obtaining X_i points or higher ($X_i = 0, 1, \ldots, m_i$) on item i for the GRM can be written as follows:

$$P^*(X_i|\theta, a_i, \delta_{X_i}) = \frac{e^{a_i(\theta - \delta_{X_i})}}{1 + e^{a_i(\theta - \delta_{X_i})}}, \tag{6.4}$$

where θ is the latent trait, a_i is the discrimination parameter for item i, δ_{X_i} is the category boundary location for the category X_i (similar to the category threshold parameter in the previous models), and $P^*(X_i|\theta, a_i, \delta_{X_i})$ is the probability of a person obtaining the score of X_i or higher (De Ayala, 2013). As mentioned earlier, GRM splits a polytomous item into a series of dichotomous items using the cumulative probabilities. That is, item i consists of m_i dichotomous items that share the same discrimination parameter (a_i) but have unique difficulty parameters (δ_{X_i}).

In the following example, we estimate item parameters for the rse data set using GRM. We use itemtype = "graded" to fit the GRM.

```
grm_mod <- "selfesteem = 1-10"
grm_fit <- mirt(data = rse[, 1:10], model = grm_mod,
            itemtype = "graded", SE = TRUE)
grm_params <- coef(grm_fit, IRTpars = TRUE, simplify = TRUE)
```

Next, we save the estimated item parameters as **grm_items** and print them.

```
grm_items <- as.data.frame(grm_params$items)
grm_items
```

```
          a          b1          b2         b3
Q1  2.324817 -1.9096256 -0.86459285 0.6212696
Q2  2.136179 -2.1468087 -1.15272898 0.6594750
Q3  2.435460 -1.3941348 -0.25807123 1.0778067
Q4  1.650816 -2.4179124 -0.86144982 1.2003060
Q5  2.172579 -1.2073811 -0.05750731 1.0257647
Q6  3.202812 -1.2249932 -0.08533037 1.1446882
Q7  2.810958 -1.0605179  0.09078729 1.3417869
Q8  1.420458 -1.2018548  0.51592382 1.7199115
Q9  2.163341 -0.7728069  0.63992017 1.4899442
Q10 2.645276 -0.8719370  0.23061101 0.9486129
```

In the output, a1 is the item discrimination parameter and the remaining columns (i.e., b1 through b3) represent the category boundary locations for the items in the **rse** data set. Similar to the output from GPCM, each item has a unique discrimination parameter. Like the GRPM, Q6 and Q8 have the highest and lowest discrimination parameters in the **rse** data set.

Next, we draw the OCC plots and examine the slopes of the option characteristic curves for Q5 and Q9, which have similar discrimination levels but differ by their category boundary locations. Figure 6.8 shows that the slopes of the OCCs for Q5 and Q9 are very similar, although Q5 appears to be easier than Q9.

```
plot(grm_fit, type = "trace", which.items = c(5, 9),
    par.settings = simpleTheme(lty = 1:4,lwd = 2),
    auto.key = list(points = FALSE, lines = TRUE, columns = 4))
```

Finally, as we have done for the previous models, we can draw the plots of TIF and cSEM for the GRM by using **type = "info"** for the TIF plot and **type = "SE"** for the cSEM plot.

```
plot(grm_fit, type = "info", theta_lim = c(-6, 6))
plot(grm_fit, type = "SE", theta_lim = c(-6, 6))
```

In addition to the traditional GRM, the **mirt** function is also capable of estimating a hybrid model that combines certain characteristics of the RSM and the GRM into a single model. This model is called the rating-scale graded

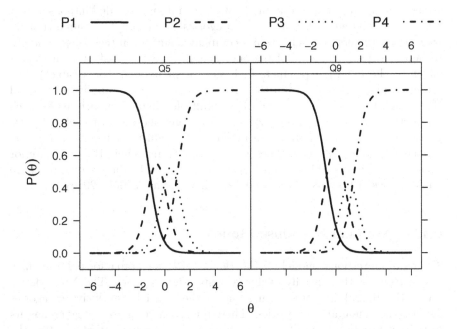

FIGURE 6.8
Option characteristic curves for Q5 and Q9 for GRM.

response model in the `mirt` package manual. To estimate this hybrid model, `itemtype` in the `mirt` function should be set to either `itemtype = "grsm"` or `itemtype = "grsmIRT"`. More details about the rating-scale graded response model can be found in Muraki (1990).

6.4 Polytomous IRT Models for Nominal Items

The previous two sections focused on Rasch and non-Rasch IRT models for polytomous items with ordered response categories. These models assume that the response categories of a polytomous item are inherently ordered. When the response categories of a polytomous item do not have an ordinal relationship, we cannot assume an ordinal transition from one response category to another as a function of the latent trait. The items with distinct response categories that cannot be ordered are called nominal items (see Chapter 2 for more information about nominal scales). Nominal IRT models directly estimate the probability of selecting a particular response category against all other categories in a nominally scored item. That is, each item is divided into a series of binary items based on the nominal categories. For example, assume that we give the customers of an electronics company a survey on technology and entertainment products. One of the questions in the survey asks the customers which of the following products (a laptop, a smartphone, or a tablet) they are most likely to buy within the next six months. If we applied a nominal IRT model to this item, we would have multiple binary comparisons for each product — such as a laptop versus either a smartphone or a tablet; a smartphone versus either a laptop or a tablet; and a tablet versus either a laptop or a smartphone. Although there are several IRT models in the literature for such nominal items, this section focuses on two models, the nominal response model (Bock, 1972) and the nested logit model (Suh & Bolt, 2010).

6.4.1   Nominal Response Model

The nominal response model (NRM; Bock, 1972) is suitable for polytomously scored items with mutually exclusive response categories. The NRM determines the probability of selecting one of the mutually exclusive categories against the remaining categories. Therefore, each response category has its own discrimination and difficulty parameters. The probability of selecting the kth response category $(k = 0, 1, \ldots, m)$ on item i in the NRM can be written as:

$$P(X_{ik}|\theta, \mathbf{a}, \boldsymbol{\gamma}) = \frac{e^{\gamma_{ik}+a_{ik}\theta}}{\sum_{h=1}^{m} e^{\gamma_{ih}+a_{ih}\theta}}, \tag{6.5}$$

where $\mathbf{a}$ is the vector of item discrimination parameters ($\mathbf{a} = [a_{i1}, a_{i2}, \ldots, a_{im}]$) for item i, $\boldsymbol{\gamma}$ is the vector of difficulty parameters ($\boldsymbol{\gamma} = [\gamma_{i1}, \gamma_{i2}, \ldots, \gamma_{im}]$) for item i, and m is the number of response categories in item i.

In the NRM, there are two model constraints regarding the item parameters to ensure model identification: (1) the sum of the item discrimination parameters is equal to zero (i.e., $\Sigma_{k=1}^{m} a_{ik} = 0$), and (2) the sum of the item difficulty parameters is equal to zero (i.e., $\Sigma_{k=1}^{m} \gamma_{ik} = 0$). It should be noted that the item difficulty parameter in NRM does not have the same meaning as item difficulty parameters in dichotomous IRT models or step parameters in polytomous IRT models for ordinal items. De Ayala (2013) describes the item difficulty parameter in NRM as the respondent's propensity to select a particular response category (e.g., continuing from the example above, selecting a smartphone instead of a tablet or a laptop). The larger the a_k value, the more strongly the response category k is related to the latent trait being measured. For example, in the context of a multiple-choice item with several response options (e.g., A, B, C, and D), the correct response option is expected to have a positive and high discrimination parameter since it is directly related to the latent trait.

To demonstrate how to fit NRM, we use the `VerbAggWide` data set from the **hemp** package. This data set comes from a study of verbal aggression (Vansteelandt, 2000). A sample of 316 respondents (243 females and 73 males) answered a questionnaire of 24 items. Each item presented a scenario, such as "a bus fails to stop for me" and the respondent was asked whether he or she would curse, scold, or shout and whether he or she would actually do this verbal aggression behavior or only want to do it. The combination of four scenarios with three verbal aggression behaviors (curse, scold, or shout) and two behavior types (want or do) resulted in 24 items in total. For each item, the respondents selected one of the following response categories: 0 (No), 1 (Perhaps), or 2 (Yes).

In the following example, we consider the three response options (no, perhaps, and yes) as nominal response categories and estimate the item parameters for NRM. The first three columns of the `VerbAggWide` data set consist of a respondent identifier, anger scores from a different measure, and gender. Therefore, we select the remaining columns by specifying `VerbAggWide[, 4:27]`. We use `itemtype = "nominal"` in the `mirt` function to fit the NRM. Then we save the results as `nrm_fit`, extract the estimated parameters into `nrm_params`, reformat the item parameters into a separate data frame (`nrm_items`), and then print the first four rows of the results using the **head** function.

```
nrm_mod <- "agression = 1 - 24"
nrm_fit <- mirt(data = VerbAggWide[, 4:27], model = nrm_mod,
                itemtype = "nominal", SE = TRUE)
nrm_params <- coef(nrm_fit, IRTpars = TRUE, simplify = TRUE)
nrm_items <- as.data.frame(nrm_params$items)
head(nrm_items, 4)
```

```
                   a1            a2         a3          c1
S1WantCurse -0.7917547 -0.019926254  0.8116810 -0.2634509
S1WantScold -1.0115207 -0.065280252  1.0768009  0.1389062
S1WantShout -0.8962308  0.138970939  0.7572599  0.4749190
S2WantCurse -0.8834298 -0.003839589  0.8872693 -0.6258154
                    c2          c3
S1WantCurse 0.05862217   0.2048288
S1WantScold 0.02968214  -0.1685883
S1WantShout 0.13833928  -0.6132583
S2WantCurse 0.28483643   0.3409790
```

In the output, a1, a2, and a3 represent the discrimination parameters of the three response options (i.e., no, perhaps, and yes). The columns labeled as c1, c2, and c3 represent the difficulty parameters (i.e., γ_k in Equation 6.5).

The results show that the third response option (i.e., yes) has the highest discrimination parameter for all items. This is an expected finding because the items in the VerbAggWide data set measure the latent trait of verbal aggression and the yes option indicates higher aggression. That is, the yes option is more capable of identifying respondents with low and high verbal aggression levels.

As described earlier, the sums of the discrimination and difficulty parameters across the three response options are constrained to be zero. We can see these constraints by summing up the parameters for the items. We use the rowSums function to sum up the discrimination and difficulty parameters and save the results into a data frame named sum_constraints. When we print the first six rows of the sum_constraints data set using the head function, we can see that the parameters are indeed summed to zero as a result of the parameterization for the NRM.

```
sum_constraints <- data.frame(
  discrimination = round(rowSums(nrm_items[, 1:3]), 3),
  difficulty = round(rowSums(nrm_items[, 4:6]), 3))
head(sum_constraints)
```

	discrimination	difficulty
S1WantCurse	0	0
S1WantScold	0	0
S1WantShout	0	0
S2WantCurse	0	0
S2WantScold	0	0
S2WantShout	0	0

Next, we draw the OCC plots for items 3 and 15 (i.e., S1WantShout and S1DoShout in the VerbAggWide data set). Figure 6.9 shows that in the context of Scenario 1 (i.e., a bus fails to stop for me), doing the action of shouting requires higher levels of verbal aggression than wanting to shout. On item 15, selecting the response option of yes (i.e., P3 in the plot) requires a higher level of verbal aggression, whereas in item 3, selecting either perhaps or yes requires relatively lower levels of verbal aggression.

```
plot(nrm_fit, type = "trace", which.items = c(3, 15),
    par.settings = simpleTheme(lty = 1:3, lwd = 2),
    auto.key = list(points = FALSE, lines = TRUE, columns = 3))
```

We can also see the OCCs separately for each response option for a particular item. Figure 6.10 shows the OCCs for item 15 in the VerbAggWide data set. We include CE = TRUE to draw the 95% confidence intervals around the OCC lines.

```
itemplot(nrm_fit, item = 15, CE = TRUE)
```

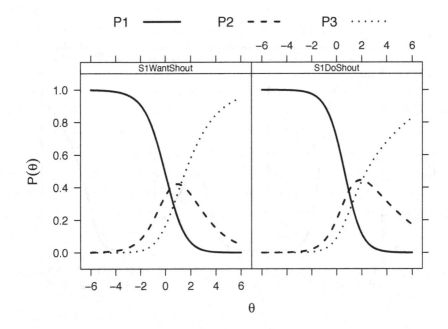

FIGURE 6.9
Option characteristic curves for items 3 and 15 for NRM.

6.4.2 Nested Logit Model

The nested logit model (NLM; Suh & Bolt, 2010) is suitable for multiple-choice items where the correct response option and all incorrect response options (i.e., distractors) are modeled separately. Therefore, the NLM can be considered a variant of the NRM that combines a dichotomous IRT model (e.g., 2PL or 3PL models) with the NRM. According to Suh and Bolt (2010), the NLM has the potential to retain the information provided by distractors when it is beneficial and to ignore the distractor information when modeling the correct response option is of main interest. A typical NLM consists of two levels. Level 1 describes the probability of selecting the correct response option versus the distractors and Level 2 deals with modeling the probabilities associated with each distractor.

Although there are several variants of the NLM, we begin with the 3PL version of NLM (3PL-NLM) because the other versions of the NLM are special cases of the 3PL-NLM with additional constraints on the discrimination and guessing parameters. The 3PL-NLM can be mathematically expressed as:

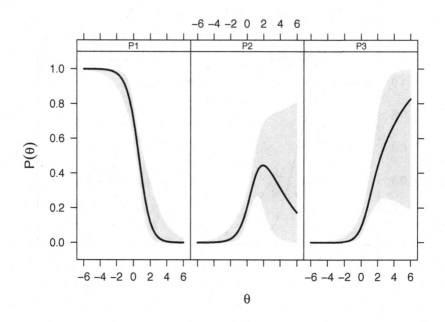

FIGURE 6.10
Option characteristic curves for item 15 for NRM.

$$P(X_i = 0, D_{iv} = 1|\theta) = P(X_i = 0|\theta)P(D_{iv} = 1|X_i = 0, \theta)$$

$$= \left\{1 - \left[c_i + (1 - c_i)\frac{1}{1 + exp^{-(b_i + a_i\theta)}}\right]\right\}\left[\frac{exp^{Z_{iv}(\theta)}}{\sum_{k=1}^{m} exp^{Z_{ik}(\theta)}}\right], \quad (6.6)$$

where $P(X_i = 0, D_{iv} = 1|\theta)$ is the probability that an examinee answers item i incorrectly and selects distractor category v given the latent trait θ, c_i is the guessing parameter for item i, b_i is the difficulty parameter for item i, a_i is the discrimination parameter for item i, and $Z_{ik}(\theta) = \gamma_{iv} + a_{iv}\theta$, as in the numerator of Equation 6.5. In the 3PL-NLM, the first part of Equation 6.6 models the correct response probability and the second part of Equation 6.6 models the individual probabilities for selecting each distractor.

The 2PL-NLM can be defined as a special form of the 3PL-NLM where the guessing parameter for item i (i.e., c_i) is constrained to be zero. Therefore, the 2PL-NLM can be written as follows:

$$P(X_i = 0, D_{iv} = 1|\theta) = \left\{1 - \left[\frac{1}{1 + exp^{-(b_i + a_i\theta)}}\right]\right\}\left[\frac{exp^{Z_{iv}(\theta)}}{\sum_{k=1}^{m} exp^{Z_{ik}(\theta)}}\right]. \quad (6.7)$$

To demonstrate how to fit the 2PL-NLM and 3PL-NLM, we use the multiplechoice data set from the **hemp** package. This data set contains item responses from a hypothetical test that consists of 27 multiple-choice items administered to 496 examinees. Each item has four response options (A, B, C, and D) but the response options are numerically represented in the data set (i.e., A = 1; B = 2; C = 3; and D = 4). In the following examples, we consider the four response options as nominal categories, and thus provide the answer key to define the correct response option for the items.

We begin the example with the 2PL-NLM. We use itemtype = "2PLNRM" in the **mirt** function to select the 2PL-NLM for estimating the item parameters. We then extract the estimated item parameters, save them to twoplnlm_items, and then print the first six rows using the **head** function.

```
key = c(4, 3, 2, 3, 4, 3, 2, 3, 1,
        4, 3, 2, 3, 3, 4, 2, 4, 3,
        3, 2, 2, 1, 2, 1, 1, 2, 1)
twoplnlm_mod <- "ability = 1 - 27"
twoplnlm_fit <- mirt(data = multiplechoice,
                model = twoplnlm_mod, itemtype = "2PLNRM",
                SE = TRUE, key = key)
twoplnlm_params <- coef(twoplnlm_fit, IRTpars = TRUE,
                simplify = TRUE)
twoplnlm_items <- as.data.frame(twoplnlm_params$items)
head(twoplnlm_items)
```

```
              a          b g u           a1          a2
item1 0.5235573 -3.4086260 0 1  -0.4839230 -0.59603034
item2 0.4940809 -2.8496901 0 1   0.2270783 -0.35252767
item3 0.5226378 -0.5159370 0 1   0.1343142 -0.05036814
item4 0.6871739 -1.6939763 0 1   0.2516607  0.29305060
item5 0.2097941  2.9218435 0 1   0.2643272 -0.03633783
item6 0.5709976 -0.4476874 0 1   0.1282541 -0.41405909
              a3         c1          c2          c3
item1  1.07995330 -0.2490648 -0.05884263  0.3079074
item2  0.12544933  0.6741845 -0.04293720 -0.6312473
item3 -0.08394604  0.4162664 -0.79558238  0.3793160
item4 -0.54471132 -0.6398626  0.66367996 -0.0238174
item5 -0.22798933  0.8475738  0.50175068 -1.3493244
item6  0.28580500 -0.4715466  0.22261253  0.2489341
```

In the output, the first few columns (a, b, g, and u) are the typical item parameters for the 2PL model described in Chapter 5. They indicate the discrimination, difficulty, lower asymptote (i.e., guessing), and upper asymptote parameters. Because the 2PL model does not involve the pseudo-guessing and upper asymptote parameters, they are fixed to 0 and 1 for all items, respec-

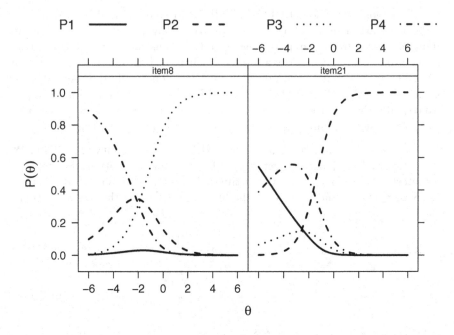

FIGURE 6.11
Option characteristic curves for items 8 and 21 for 2PL-NLM.

tively. The following columns in the output (a1, a2, a3, c1, c2, and c3) are NRM-based parameters for the three distractors in the multiple-choice items.

In Figure 6.11, we see the OCCs for items 8 and 21 based on the estimated parameters from the 2PL-NLM. The third response category (P3) for item 8 and the second response category (P2) for item 21 have positive slopes because these are the correct response categories for items 8 and 21. The curves for these categories start around $P(\theta) = 0$ because the guessing parameter is equal to zero in the 2PL-NLM. For both items, the fourth response option D labeled as P4 appears to be the most plausible distractor for low-ability examinees. Especially in item 8, there is a very high likelihood that a low-ability examinee (e.g., $\theta < -4$) would choose the fourth response option, which is a distractor in this item. However, given the nearly flat curve for the first response option, P1, for item 8, this is the least plausible distractor. That is, nearly none of the examinees would choose this particular distractor. A similar interpretation can be made for the third response option, P3, in item 21.

```
plot(twoplnlm_fit, type = "trace", which.items = c(8, 21),
    par.settings = simpleTheme(lty = 1:4, lwd = 2),
    auto.key = list(points = FALSE, lines = TRUE, columns = 4))
```

Next, we use the same data set to fit the 3PL-NLM. This time, we use itemtype = "3PLNRM" to estimate the item parameters based on the 3PL-NLM, save the estimated item parameters in threeplnlm_items, and again print the first few rows with the head function.

```
threeplnlm_mod <- "ability = 1 - 27"
threeplnlm_fit <- mirt(data = multiplechoice,
                model = threeplnlm_mod, itemtype = "3PLNRM",
                SE = TRUE, key = key)
threeplnlm_params <- coef(threeplnlm_fit, IRTpars = TRUE,
                simplify = TRUE)
threeplnlm_items <- as.data.frame(threeplnlm_params$items)
head(threeplnlm_items)
```

	a	b	g	u	a1	a2	a3
item1	0.6874	-1.0233	0.5512	1	-0.2967	-0.6796	0.9763
item2	0.5021	-2.7623	0.0176	1	0.2133	-0.3827	0.1693
item3	0.5759	-0.3280	0.0408	1	0.1494	-0.0432	-0.1062
item4	0.6464	-1.7032	0.0334	1	0.1639	0.3439	-0.5078
item5	0.2192	2.8837	0.0064	1	0.2337	-0.0138	-0.2199
item6	0.8618	0.6145	0.2781	1	0.1429	-0.3940	0.2511

	c1	c2	c3
item1	-0.1298	-0.1373	0.2671
item2	0.6785	-0.0668	-0.6118
item3	0.4218	-0.7924	0.3707
item4	-0.6654	0.6694	-0.0040
item5	0.8453	0.5042	-1.3495
item6	-0.4749	0.2358	0.2391

In the output, g shows the estimated guessing parameters for the items. The results show that some items in the multiplechoice data set (e.g., items 1, 7, 17, 19 and 25) have high effects of guessing. Therefore, after accounting for the effect of guessing in these items in the 3PL-NLM, the estimated item difficulty and item discrimination parameters have significantly changed compared to the item parameters from the 2PL-NLM.

To demonstrate the impact of guessing, we draw the OCCs for items 1 and 17 based on the estimated parameters from the 3PL-NLM. Figure 6.12 shows that the fourth response option (P4) is the correct response option for both item 1 and item 17. Item 1 has a higher guessing parameter than item 17. The second response option, labeled as P2, seems to be the best distractor for item 1, whereas the third response option, labeled as P3, is the best distractor for item 17. Figure 6.12 also suggests that low-ability examinees are more likely to choose the best distractor P3 in item 17 despite the high degree of guessing involved in the item.

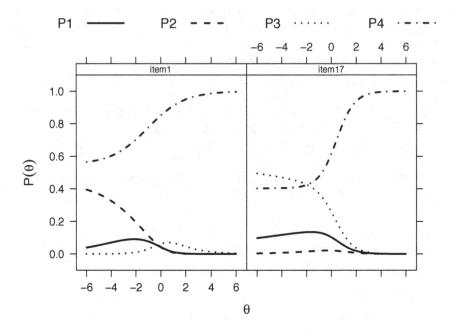

FIGURE 6.12
Option characteristic curves for items 1 and 17 for 3PL-NLM.

```
plot(threeplnlm_fit, type = "trace", which.items = c(1, 17),
     par.settings = simpleTheme(lty = 1:4, lwd = 2),
     auto.key = list(points = FALSE, lines=TRUE, columns=4))
```

6.5   Model Selection

Model selection in IRT depends on both theoretical assumptions regarding the design of an assessment as well as the conditions of item response data (i.e., item and person fit). We have described how to use the itemfit and personfit functions to examine item and person fit for the IRT models for dichotomous items in Chapter 5. The same procedures also apply to the IRT models for polytomous items demonstrated in Chapter 6.

For polytomous items with ordered response categories, one may prefer to use either the PCM or the RSM, assuming all the items have the same level of discrimination. If, however, the items are expected to differ significantly regarding their discrimination levels, either the GRM or the GPCM may be more suitable for estimating item parameters. If item response categories are nominal rather than ordinal, then the NRM can be used for analyzing the item response data.

6.6 Summary

In this chapter, we introduced the unidimensional IRT models for polytomous item responses and demonstrated how to estimate the PCM, GPCM, RSM, GRM, NRM, and NLM using the `mirt` function in the **mirt** package. The `mirt` function is capable of estimating traditional IRT models as well as their specialized forms for polytomous items. We would also like to encourage the interested reader to look into the **TAM** package (Kiefer et al., 2016), which is another comprehensive R package for estimating polytomous IRT models. The IRT models presented in this chapter are suitable for unidimensional test structures consisting of polytomous items with ordered or nominal response categories. The same graphical tools and model diagnostic options introduced in Chapter 5 are also available for the polytomous IRT models presented in this chapter. In the next chapter, we focus on multidimensional extensions of the unidimensional IRT models that we introduced in Chapters 5 and 6.

7

Multidimensional Item Response Theory

7.1 Chapter Overview

In Chapters 5 and 6, we presented IRT models for dichotomously and polytomously scored items and demonstrated how to estimate these models in R. A common characteristic of the IRT models presented in Chapters 5 and 6 is that there is a single latent trait underlying the items. In other words, the test structure is assumed to be unidimensional. However, most educational and psychological assessments tend to measure two or more latent traits or dimensions (Ackerman, Gierl, & Walker, 2003). When the items measure two or more latent traits, the test structure is referred to as multidimensional. To estimate item parameters and latent trait levels from multidimensional assessments, multidimensional IRT (MIRT) models should be used.

Chapter 7 focuses on multidimensionality and the fitting of MIRT models in R. The chapter begins with a definition of multidimensionality in the context of educational and psychological measurement, followed by a brief discussion of MIRT-related concepts (e.g., compensatory models vs. noncompensatory models, within-item vs. between-item multidimensionality, and exploratory vs. confirmatory MIRT models). Then, we demonstrate how to fit common MIRT models, specifically the multidimensional Rasch model, multidimensional 2PL model, multidimensional graded response model, and the bi-factor IRT model, using the **mirt** package (Chalmers, 2012). The **mirt** package is capable of estimating various MIRT models, although there are other R packages available for estimating some MIRT models, such as **flirt** (Jeon, Rijmen, & Rabe-Hesketh, 2014), **TAM** (Kiefer et al., 2016), **eRm** (Mair & Hatzinger, 2007a), and **sirt** (Robitzsch, 2017).

7.2   Multidimensional Item Response Modeling

As we have discussed in Chapters 5 and 6, unidimensional IRT models are appropriate for analyzing examinees' responses to items measuring a single latent trait or construct. However, for complex items that measure multiple latent traits simultaneously, item responses should not be modeled using a unidimensional IRT model because the probability of a correct response would not be a function of a single latent trait anymore. For analyzing complex items that involve multiple latent traits, a *multidimensional* IRT approach (i.e., MIRT models) should be used. MIRT models explain the probability of a correct response to a given item as a function of a vector of latent traits, instead of a single latent trait. Although MIRT can be used for various purposes, this modeling framework is particularly useful

1. for modeling the item response data when one or more items measure multiple latent traits simultaneously (Reckase, 2009), and

2. for improving measurement precision when there are individual tests or subtests measuring correlated latent traits (Wang, Chen, & Cheng, 2004).

Since the early 1980s, researchers have proposed different MIRT models (e.g., Adams, Wilson, & Wang, 1997; Reckase, 1985; Yao & Schwarz, 2006), which are often multidimensional extensions of the unidimensional IRT models. In this chapter, we focus on some MIRT models and demonstrate how to fit and estimate these models using the **mirt** package in R. We encourage our readers to review *Multidimensional Item Response Theory* by Reckase (2009) for a more comprehensive discussion of multidimensional item response modeling.

MIRT models can be categorized in several ways. For example, MIRT models can be categorized based on whether the weakness (or deficit) in one dimension (i.e., latent trait) can be compensated by the strength in another dimension. In addition, MIRT models can be grouped based on whether each item is associated with a single latent trait or multiple latent traits. In the following sections, we briefly explain these categorizations.

7.2.1   Compensatory and Noncompensatory MIRT

MIRT models can be characterized as either compensatory or noncompensatory based on the presence of a compensatory relationship between the latent traits (Sijtsma & Junker, 2006). The additive nature of *compensatory*

MIRT models defines the probability of a correct response based on the sum of a series of latent traits weighted with different slope (i.e., discrimination) parameters. Compensatory MIRT models allow the weakness in one latent trait to be compensated by the strength in another latent trait (Reckase, 1997; Yao & Boughton, 2007). For example, an examinee with a low proficiency level in arithmetic can answer a math problem correctly with the help of a higher proficiency level in algebra.

Figure 7.1 shows a graphical illustration of a dichotomous item from a two-dimensional, compensatory MIRT model. Unlike an item characteristic curve in unidimensional IRT models, the MIRT model yields an item characteristic surface because the item is associated with two latent dimensions, namely θ_1 and θ_2. For this particular item, the two dimensions have a compensatory relationship. The item surface plot shows that an examinee with a low proficiency level in dimension 1 (e.g., $\theta_1 = 0$) and a high proficiency level in dimension 2 (e.g., $\theta_2 = 2$) would still have a high probability of answering the item correctly because the low proficiency in dimension 1 is compensated for by the high proficiency in dimension 2.

Unlike the compensatory models, the multiplicative nature of *noncompensatory* MIRT models defines the probability of a correct response based on the multiplication of probabilities, each based on a different latent trait. Therefore, the weakness in one dimension cannot be compensated by the strength in another dimension. For example, an examinee with a low proficiency in arithmetic may not answer a word problem correctly because of a higher proficiency level in reading because the item requires both skills. It should be noted that noncompensatory MIRT models are also known as partially compensatory models because noncompensatory models also allow for some compensation (Reckase, 2009). Although there have been several applications of noncompensatory MIRT models in the cognitive assessment literature (e.g., Embretson, 1997; Junker & Sijtsma, 2001; Maris, 1995), compensatory MIRT models have been more popular among researchers because of the theoretical similarities between compensatory models and factor analysis models (Sijtsma & Junker, 2006).

Figure 7.2 shows an item characteristic surface for a dichotomous item from a two-dimensional, noncompensatory MIRT model. The item surface plot shows that it is essential to have a high proficiency level in both dimensions to answer the item correctly. That is, an examinee with a low proficiency in one dimension would not be able to answer the item correctly despite having a higher proficiency level in the other dimension.

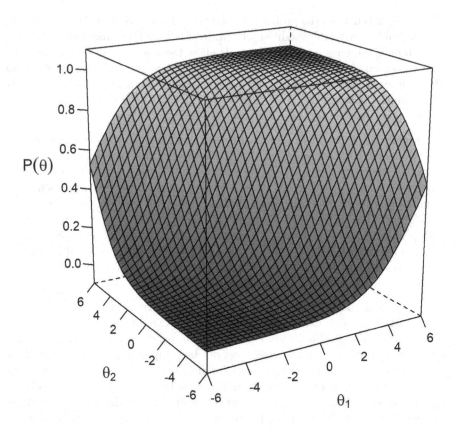

FIGURE 7.1
Item characteristic surface for a compensatory MIRT model.

7.2.2   Between-Item and Within-Item Multidimensionality

MIRT models can be categorized into two groups based on the test structure. These are between-item models and within-item MIRT models (Adams et al., 1997; Wang et al., 2004). In a between-item MIRT model, each latent trait is measured by a unique set of items. In other words, each item is associated with only one of the latent traits measured by the test. This type of test structure is also known as simple structure, as with a factor analytic solution where items load onto a single factor (see Chapter 4 for a review of factor analysis). Between-item MIRT models are typically used for estimating subscores from subtests measuring correlated latent traits (e.g., Bulut, Davison, & Rodriguez, 2017; de la Torre, Song, & Hong, 2011; Wang et al., 2004). In a within-item MIRT model, each item can be associated with two or more latent traits measured by the test. This type of test structure is known as a nonsimple or

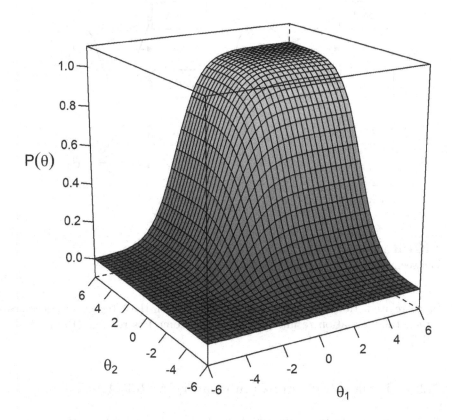

FIGURE 7.2
Item characteristic surface for a noncompensatory MIRT model.

complex structure. A good example of within-item multidimensionality is the bi-factor test structure from cognitive ability assessments, where all items are associated with a general latent trait (i.e., general intelligence) and each item is associated with a secondary latent trait (i.e., specific cognitive abilities).

Figure 7.3 illustrates the between-item and within-item MIRT models for a test measuring two latent traits. In the between-item model, item 1 to item 5 are associated with the first latent trait, θ_1, while item 6 to item 10 are associated with the second latent trait, θ_2. θ_1 and θ_2 do not share any common items. Therefore, the test structure becomes a multi-unidimensional structure where each latent trait is defined by a set of unidimensional items while the overall structure becomes multidimensional. In the within-item model, the items are associated with both θ_1 and θ_2, and thus the test has a nonsimple structure. In the MIRT models presented in Figure 7.3, θ_1 and θ_2 are likely to

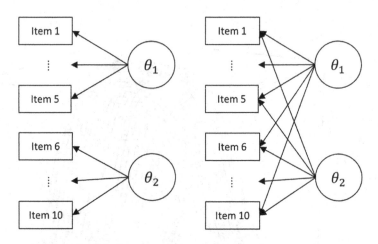

FIGURE 7.3
Between-item (left) and within-item (right) structure.

be correlated. However, as the correlation between the latent traits becomes larger, the estimation results are likely to become less reliable (Bulut et al., 2017).

7.2.3   Exploratory and Confirmatory MIRT Analysis

When analyzing response data with MIRT models, the test structure can be defined in either an exploratory or confirmatory way. If there is no a priori hypothesis for the test structure underlying the items, only the expected number of dimensions needs to be specified before estimating the item parameters and latent trait levels. This is the exploratory MIRT approach where the association between the latent traits and items are derived from the data. If there is a hypothesis regarding the test structure, then both the number of dimensions and the relationship between the items and the dimensions need to be specified (Reckase, 2009, p. 179). This is called the confirmatory MIRT approach because the test structure is merely specified by the researcher based on theory. The exploratory MIRT analysis is often used for finding the number of dimensions underlying the data. This type of analysis is analogous to exploratory factor analysis, which we discussed in Chapter 4. In this chapter, we focus on the confirmatory MIRT models where we define the number of dimensions as well as the relationship between items and latent traits (analogous to confirmatory factor analysis, also presented in Chapter 4).

7.3 Common MIRT Models

In this chapter, we highlight some of the common MIRT models. These models include the multidimensional 2PL model, the multidimensional Rasch model, the multidimensional graded response model, and the bi-factor IRT model. In the following sections, we explain these MIRT models briefly and demonstrate how to fit and estimate these models using the **mirt** package.

7.3.1 Multidimensional 2PL Model

The multidimensional 2PL (M2PL) model is an extension of the unidimensional 2PL model. The M2PL model is a compensatory MIRT model, capable of estimating item difficulty and discrimination parameters from a multidimensional test. The test structure may be either between-item or within-item. The M2PL model can be written as:

$$P(X_{ij} = 1 | \boldsymbol{\theta}_j, \mathbf{a_i}, d_i) = \frac{\exp(\mathbf{a_i}\boldsymbol{\theta}_j{}' + d_i)}{1 + \exp(\mathbf{a_i}\boldsymbol{\theta}_j{}' + d_i)}, \tag{7.1}$$

where $\boldsymbol{\theta}_j$ is a $1 \times M$ vector of latent traits for examinee j $(\boldsymbol{\theta}_j = (\theta_{j1}, \theta_{j2}, \ldots, \theta_{jM}))$, $\mathbf{a_i}$ is a $1 \times M$ vector of slope (i.e., item discrimination) parameters for item i $(\mathbf{a_i} = (a_{i1}, a_{i2}, \ldots, a_{iM}))$, and d_i is the intercept term for item i. The intercept parameter d_i is not similar to the item difficulty parameter in a unidimensional IRT model because this parameter cannot be considered a unique indicator of the difficulty of the item (Reckase, 2009). A transformation can be performed to obtain the traditional difficulty parameters from the intercept parameters. The formula for this transformation is shown below:

$$B_i = \frac{-d_i}{\sqrt{\sum_{m=1}^{M} a_{im}^2}}, \tag{7.2}$$

where B_i is the multidimensional difficulty parameter for item i, which is also referred to as MDIFF. The higher the value of B_i gets, the more difficult the item becomes.

The item discrimination parameters from the MIRT models can also be summarized with a single value using a transformation. In fact, this transformation is the denominator of Equation 7.2 as

$$A_i = \sqrt{\sum_{m=1}^{M} a_{im}^2},$$
(7.3)

where A_i is the multidimensional discrimination parameter, which is also known as MDISC in the literature. The multidimensional discrimination parameter is similar to the item discrimination parameter in the unidimensional IRT model. This parameter represents the overall discrimination level of a multidimensional item. If the item is only associated with a single latent trait, a_i in Equation 7.1 would have only one non-zero element (e.g., $a_i = (1.4, 0, 0)$ in a three-dimensional test). In this case, the multidimensional discrimination parameter would be equal to the non-zero element of a_i.

To demonstrate how to fit and estimate the M2PL model, we use the mimic data set from the **hemp** package. Before we begin the analysis, we first activate the **hemp** and **mirt** packages with the `library` command.

```
library("hemp")
library("mirt")
```

The mimic data set contains the responses from 2000 examinees to 24 items. The items are scored dichotomously (i.e., 1 = correct; 0 = incorrect). The items are labeled as item1, item2, ..., item24. The test structure is within-item multidimensionality (complex), where item1 to item6, item13 to item21, item23, and item24 are associated with the first latent trait, while item7 to item20, and item22 to item24 are associated with the second latent trait.

To fit the M2PL model, we again use the `mirt` function in the **mirt** package, as we have done for estimating the unidimensional IRT models in Chapters 5 and 6. The R commands for estimating the multidimensional IRT models are very similar to those presented in Chapters 5 and 6. The major difference is the definition of the test structure because the multidimensional test structures involve two or more latent traits rather than a single latent trait.

Because the mimic data set has a two-dimensional structure, we define two latent traits: F1 and F2. When defining the test structure, we specify the items associated with each latent trait. Instead of listing each item one by one, we can specify a range of items for each latent trait (as we did in Chapters 5 and 6). For example, 1-6 refers to item1 through item6. We separate the sets of items using a comma (i.e., 1-6, 13-21, 23-24). Because the mimic data set has a within-item test structure, item13 through item20, item23, and item24 appear for both latent traits. The final line of our model definition COV = F1 * F2 defines the covariance term that we want to estimate. In this example, we want to estimate the covariance between the two latent traits. Therefore, we include COV = F1 * F2 in the model definition. Skipping this line would fix the covariance between F1 and F2 to zero in the model, which would assume the latent traits are orthogonal. By default, the `mirt` function fixes the variances of the latent traits F1 and F2 to 1. To estimate these variances in the model, the current cov statement can be replaced with `COV = F1 * F2, F1 * F1, F2 * F2`. In addition, the means of the latent traits are fixed to zero. If the means need to be estimated, this needs to be specified in the test structure (e.g., `MEAN = F1, F2`). These constraints are made for identification reasons as we have described in the context of factor analysis in Chapter 4. We save this test structure as m2pl_mod.

```
m2pl_mod <- 'F1 = 1 - 6, 13 - 21, 23 - 24
             F2 = 7 - 20, 22 - 24
             COV = F1 * F2'
```

In the `mirt` function, we specify the model that we want to fit with `model = mod2pl_mod` and the item type with `itemtype = "2PL"`. The default estimation algorithm in the `mirt` function is `method = "EM"`, which is the expectation maximization algorithm. For the MIRT models with up to three dimensions, the EM algorithm is considered effective, but the other estimation algorithms (e.g., `method = "MHRM"`) are recommended for test structures involving more than three dimensions (see the **mirt** package manual for more details). We save the fitted model as m2pl_fit, extract the estimated parameters using the `coef` function, and save them as m2pl_params.

```
m2pl_fit <- mirt(data = mimic, model = m2pl_mod,
                 itemtype = "2PL", method = "EM",
                 SE = TRUE)
m2pl_params <- coef(m2pl_fit, simplify = TRUE)
```

The estimated item parameters can also be printed directly by adding `$items` at the end of m2pl_params. In previous chapters, we have saved the estimated item parameters as separate data sets, but here we print them directly from m2pl_params.

```
head(m2pl_params$items)
```

	a1	a2	d	g	u
item1	1.0452	0	0.0301	0	1
item2	0.9550	0	-0.3060	0	1
item3	1.0844	0	0.2320	0	1
item4	1.2282	0	0.2081	0	1
item5	0.9071	0	0.1661	0	1
item6	0.8660	0	0.7703	0	1

The output is very similar to the output that we have seen for the unidimensional IRT models. However, instead of only one discrimination parameter, there are two columns, a1 and a2, representing the discrimination (i.e., slope) parameters for the first and second latent traits, respectively. Notice that some discrimination parameters are equal to zero. It is because some items are only associated with the first or second latent trait, not both. For example, item1 has a non-zero item discrimination for a1 and 0 for a2. This indicates that item1 is associated with the first latent trait (i.e., F1), but not with the second latent trait (i.e., F2). Unlike item1, item13 has a non-zero item discrimination parameter for both a1 and a2 because this particular item is associated with both F1 and F2. The following column, d, represents the estimated intercept parameters for the items. The final two columns, g and u, are again the lower asymptote (i.e., guessing) and upper asymptote parameters. Because we used the M2PL model, the lower asymptote parameter is fixed to zero and the upper asymptote parameter is fixed to 1 for all items.

Next, we use the `MDIFF` and `MDISC` functions from the **mirt** package to transform the discrimination and intercept parameters into the multidimensional item difficulty and discrimination parameters based on Equations 7.2 and 7.3. We then combine the transformed parameters into a data frame, using the `data.frame` function, and save it as `m2pl_items`. Finally, we rename the multidimensional item parameters and print the first six items using the `head` function.

```
m2pl_items <- data.frame(MDISC(m2pl_fit),
                         MDIFF(m2pl_fit))
colnames(m2pl_items) <- c("m2pl_mdisc", "m2pl_mdiff")
head(m2pl_items)
```

```
      m2pl_mdisc  m2pl_mdiff
item1  1.0451546 -0.02879262
item2  0.9549557  0.32042841
item3  1.0843917 -0.21399037
item4  1.2282254 -0.16942062
item5  0.9071266 -0.18311061
item6  0.8660056 -0.88950791
```

In addition to item parameters, we can also see the estimated variance-covariance matrix of the two latent traits by adding `$cov` at the end of `m2pl_params`. The output shows a 2 × 2 matrix where the diagonal elements are the variances of the latent traits, F1 and F2, and the lower off-diagonal element is the covariance of the two latent traits. In the output below, the variances of F1 and F2 are 1, for the identification reasons mentioned above, and the covariance of the two latent traits is estimated at 0.588.

```
m2pl_params$cov
```

```
         F1 F2
F1 1.000000 NA
F2 0.588067  1
```

Because the variances of the latent traits are fixed to 1, the variance-covariance matrix is actually the correlation matrix of the latent traits. Therefore, the correlation between F1 and F2 also becomes .588. However, if the variances of F1 and F2 were estimated in the model, we could use the following formula to transform the estimated covariance into a correlation coefficient:

$$r_{F1,F2} = \frac{cov(F1, F1)}{(S_{F1})(S_{F2})} = \frac{0.588}{(\sqrt{1})(\sqrt{1})} = .588, \tag{7.4}$$

where S_{F1} and S_{F2} are the standard deviations of the latent traits F1 and F2 and $cov(F1, F1)$ represents the covariance between the two latent traits.

As we have done for the unidimensional IRT models, we can graphically examine the item and test characteristics for the MIRT models. The MIRT models yield item and test surface plots where multiple latent traits are used together to calculate the probabilities of responding correctly to an item, to calculate item and test information, and to examine other properties. There are also contour plots capable of conveying the same information as the surface plots. A contour plot presents the information from a bird's-eye view. In other words, the contour plot shows the surface plot as if it were seen from above. The contour plots are often easier to interpret relative to the surface plots. It should be noted that the MIRT models involving more than two latent traits are harder to summarize graphically regardless of which type of plot is chosen because (1) the plot functions in the **mirt** package can handle only two dimensions; and (2) the other graphical functions in R are limited to data from two-dimensional and three-dimensional space.

The following example again uses the `itemplot` function to plot the items. We use item13 in the examples below because this particular item is one of the items associated with both latent traits. Using the `itemplot` function with the `type = "trace"` option, we first create an item characteristic surface plot for item13. This is a three-dimensional plot with the following axes: the first latent trait on the x axis, the second latent trait on the y axis, and the probability of answering item13 correctly on the z axis. Then, we create an item contour plot for the same item using `type = "tracecontour"`.

```
itemplot(m2pl_fit, type = "trace", item = 13)
itemplot(m2pl_fit, type = "tracecontour", item = 13)
```

Both Figures 7.4 and 7.5 show how the probability of answering item13 correctly changes as a function of the first latent trait (θ_1) and the second latent trait (θ_2). Because the M2PL model is a compensatory MIRT model, the weakness in one latent trait can be compensated by the strength in the other latent trait. For example, for $\theta_1 = 0$ and $\theta_2 = 2$, the probability of answering item13 correctly is nearly .80 (i.e., 80%) despite θ_1 being relatively lower than θ_2.

In addition to the item characteristic surface, we can also examine the information provided by the item based on the latent trait levels. To create

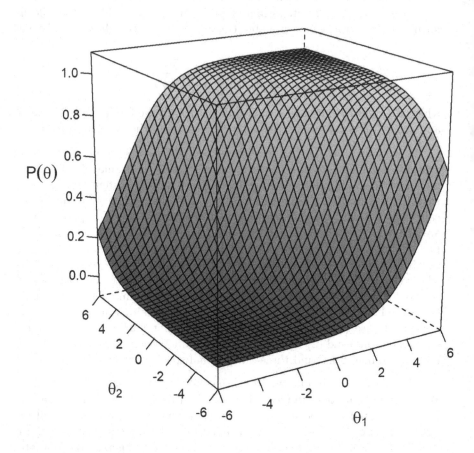

FIGURE 7.4
Item surface plot for item13 for the M2PL model.

an item information plot, we set `type = "info"` in the `itemplot` function. The resulting plot is similar to the item characteristic surface plot, but this time the z-axis shows the item information level instead of the probability of success in the item. Figure 7.6 shows that the item information level is the highest when both latent traits are around zero, whereas the item information is the lowest when both latent traits become either very low or very high.

```
itemplot(m2pl_fit, type="info", item = 13)
```

Like the item characteristic surface plot, the item information plot can also be drawn as a contour plot. In addition, the `itemplot` function can be used

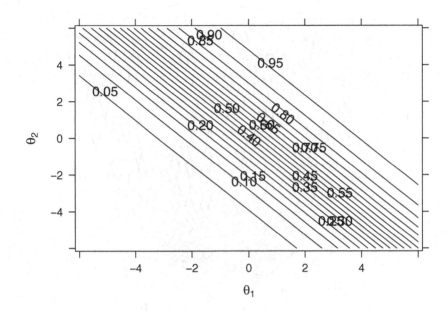

FIGURE 7.5
Item contour plot for item13 for the M2PL model.

for examining the expected score and standard error values for the items. The R commands for drawing these plots are shown below.

```
itemplot(m2pl_fit, type = "infocontour", item = 13)
itemplot(m2pl_fit, type = "score", item = 13)
itemplot(m2pl_fit, type = "SE", item = 13)
```

In addition to the item-level plots summarized above, it is also possible to draw plots at the test level. These plots summarize the test information function and conditional standard error of measurement. The following commands show the use of the `plot` function for creating a TIF plot and a cSEM plot. Figure 7.7 shows the TIF and cSEM plots for all items included in the `mimic` data set.

```
plot(m2pl_fit, type = "info")
plot(m2pl_fit, type = "SE")
```

In Chapter 5, we demonstrated how to use the `fscores` function in the **mirt** package for estimating the latent trait levels. For the MIRT models, we can follow the same procedure to estimate the latent trait levels from a multidimensional assessment. In the example below, we use the maximum a

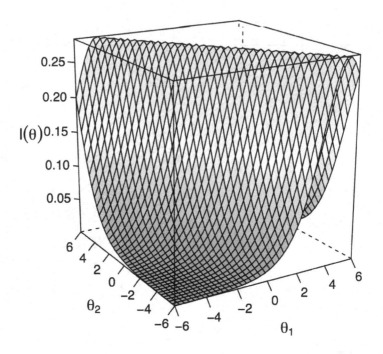

FIGURE 7.6
Item information function plot for item13 for the M2PL model.

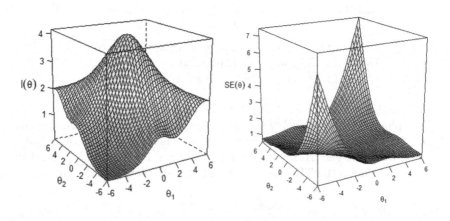

FIGURE 7.7
Test information function (left) and cSEM (right) plots for the M2PL.

posteriori (MAP) and expected a posteriori (EAP) methods to estimate the latent trait levels from the M2PL model. We specify `method = "MAP"` and method = "EAP" for the MAP and EAP estimators, respectively. The ML-based estimates can also be obtained using method= " ML" in the fscores function. As we have explained in Chapter 5, the maximum likelihood (ML) estimation fails to provide a result when all item responses are either incorrect or correct for an examinee. The same issue persists when estimating the latent trait levels from the MIRT models. Below we estimate the latent trait levels, save the results as m2pl_map and m2pl_eap, and then print the first few rows with the **head** function.

```
m2pl_map <- fscores(m2pl_fit, method = "MAP", full.scores = TRUE,
                    full.scores.SE = TRUE)
head(m2pl_map)
```

	F1	F2	SE_F1	SE_F2
[1,]	0.4244960	0.03977301	0.4836689	0.4734518
[2,]	0.8752283	1.29303637	0.5318044	0.5312877
[3,]	1.1437356	0.51740865	0.5330358	0.4934871
[4,]	-0.2365629	-0.45059857	0.4752881	0.4787024
[5,]	0.9421421	-0.12299459	0.5066176	0.4782289
[6,]	0.3985295	1.09251385	0.4975982	0.5062714

```
m2pl_eap <- fscores(m2pl_fit, method =  "EAP",
                    full.scores = TRUE,
                    full.scores.SE = TRUE)
head(m2pl_eap)
```

	F1	F2	SE_F1	SE_F2
[1,]	0.4453246	0.0453828	0.4918731	0.4803416
[2,]	0.9311577	1.3492089	0.5424382	0.5421459
[3,]	1.1960052	0.5477967	0.5417752	0.5022786
[4,]	-0.2446835	-0.4667777	0.4834888	0.4855015
[5,]	0.9780828	-0.1158884	0.5144185	0.4850905
[6,]	0.4324385	1.1331393	0.5072794	0.5155978

In the output, the columns F1 and F2 are the latent trait estimates and the other columns SE_F1 and SE_F2 are the standard errors for the estimated latent traits. Next, we combine the latent trait estimates from the MAP and EAP methods in a new data set called m2pl_scores and then look at the correlations among the latent trait estimates. The results show that the latent trait estimates from the MAP and EAP methods are highly correlated. Also, we find the scores between the two latent traits are highly correlated (approximately 0.78 for the MAP and EAP).

```
m2pl_scores <- data.frame(map1 = m2pl_map[, 1],
                          map2 = m2pl_map[, 2],
                          eap1 = m2pl_eap[, 1],
                          eap2 = m2pl_eap[, 2])
```

```
cor(m2pl_scores)
```

```
          map1    map2    eap1    eap2
map1 1.0000 0.7789 1.0000 0.7839
map2 0.7789 1.0000 0.7841 1.0000
eap1 1.0000 0.7841 1.0000 0.7891
eap2 0.7839 1.0000 0.7891 1.0000
```

Depending on the theoretical assumptions and expectations about the items and the test structure, it is also possible to modify the definition of the test structure to estimate more specific MIRT models. For example, some items can be constrained to have the same slope and intercept parameters. Assuming that we expect the unique items between the two latent traits to have the same slope parameters, we can use the CONSTRAIN statement in the model definition.

```
m2pl_mod_constrant <- '
          F1 = 1 - 6, 13 - 21, 23 - 24
          F2 = 7 - 20, 22 - 24
          COV = F1 * F2
          CONSTRAIN = (1 - 6, 21, a1), (7 - 12, 22, a2)'
```

There is also a noncompensatory version of the M2PL model. To estimate the noncompensatory M2PL, we need to replace itemtype = "2PL" with itemtype = "PC2PL". Unlike in the compensatory M2PL model, the items would have a separate slope and intercept parameter for each of the two latent traits. Also, there would be no compensation between the the latent traits in the noncompensatory M2PL model.

```
m2pl_fit <- mirt(data = mimic, model = m2pl_mod,
                 itemtype = "PC2PL", SE = TRUE)
```

In the mimic data set, the items do not involve any guessing. However, if the item responses were suspected to be influenced by guessing, then the multidimensional 3PL model could be selected using itemtype = "3PL". Alternatively, a fixed lower asymptote value could be defined for all items by setting the guess parameter in the mirt function (e.g., guess = 0.10).

7.3.2 Multidimensional Rasch Model

Adams et al. (1997) introduced the multidimensional form of the Rasch model, which is capable of estimating item parameters from a test measuring multiple latent traits. This model is also known as the multidimensional random coefficients multinomial logit model in the literature. The model described by Adams et al. (1997) is a general form of the Rasch model and is capable of handling both dichotomous and polytomous items. Using the notation from Adams et al. (1997), the multidimensional Rasch model defines examinee j's probability of selecting the response category k ($k = 0, 1, 2, \ldots, K$) in item i of an M-dimensional test as

$$P(X_{ijk} = 1|\xi, \theta_j) = \frac{\exp(\mathbf{b_{ik}}'\theta_j + \mathbf{a_{ik}}'\xi)}{\sum_{k=0}^{K} \exp(\mathbf{b_{ik}}'\theta_j + \mathbf{a_{ik}}'\xi)}, \qquad (7.5)$$

where $\mathbf{b_{ik}}$ is a score vector for score category k of item i across the M dimensions, $\mathbf{a_{ik}}$ is a design vector for score category k of item i that defines the relationship between the item difficulty parameters, θ_j is a vector of latent traits $(\theta_j = (\theta_{j1}, \theta_{j2}, \ldots, \theta_{jM}))$, and ξ is a vector of item difficulty parameters for item i. It should be noted that $\mathbf{a_{ik}}$ and $\mathbf{b_{ik}}$ do not refer to item discrimination and item difficulty parameters; rather, they are weights derived from the test structure based on our hypothesis. In Equation 7.5, the only estimated parameters are ξ and θ_j.

The multidimensional Rasch model can be used for both between- and within-item test structures measuring multiple latent traits. In the multidimensional Rasch model, the correlations between latent traits serve as additional information. Through the correlations among the latent traits, precision can be improved substantially, especially when tests are short and the number of estimated latent traits is large (Wang et al., 2004). In the following example, we use the `mimic` data set to demonstrate how to fit the multidimensional Rasch model. The model definition is the same as the one from the M2PL model because the expected test structure does not change between the models. To estimate the multidimensional Rasch model, we just specify `itemtype = "Rasch"`. The rest of the elements are the same as the previous example with the M2PL model.

```
mrasch_mod <- 'F1 = 1 - 6, 13 - 21, 23 - 24
               F2 = 7 - 20, 22 - 24
               COV = F1 * F2'
mrasch_fit <- mirt(data = mimic, model = mrasch_mod,
                   itemtype = "Rasch", SE = TRUE)
mrasch_params <- coef(mrasch_fit, simplify = TRUE)
```

Next, we print the first ten rows of the estimated item parameters using the **head** function. The output is very similar to the output we have seen for the M2PL model. However, the item discrimination parameters are fixed to 1 for the items in the multidimensional Rasch model.

```
head(mrasch_params$items, 10)
```

	a1	a2	d	g	u
item1	1	0	0.02941397	0	1
item2	1	0	-0.29423817	0	1

```
item3   1  0  0.21644013 0 1
item4   1  0  0.18630579 0 1
item5   1  0  0.16316568 0 1
item6   1  0  0.76296144 0 1
item7   0  1  0.02323546 0 1
item8   0  1 -0.91000729 0 1
item9   0  1  0.53989389 0 1
item10  0  1  0.88894243 0 1
```

Next, we use the `MDIFF` function to transform the intercept parameters into the item difficulty parameters. From the output, we see that for the items associated with a single latent trait, only the sign of the d parameter changed; whereas for the items associated with both latent traits, the item difficulty parameters are different after the transformation. Based on the multidimensional Rasch model, item10 is the easiest item and item11 (not printed) is the most difficult item.

```
mrasch_mdiff <- MDIFF(mrasch_fit)
head(mrasch_mdiff)
```

```
            MDIFF_1
item1 -0.02941397
item2  0.29423817
item3 -0.21644013
item4 -0.18630579
item5 -0.16316568
item6 -0.76296144
```

Finally, we print the estimated variance-covariance matrix of the two latent traits. The output shows that the variances of F1 and F2 are 0.689 and 0.739, respectively. The covariance of the two latent traits is 0.344.

```
mrasch_params$cov
```

```
          F1        F2
F1 0.6893974        NA
F2 0.3441965 0.739076
```

As we have done for the M2PL model, we can examine the item and test characteristics visually using the `itemplot` and `plot` functions. The R commands for these plots are presented below.

```
itemplot(mrasch_fit, type = "trace", item = 13)
itemplot(mrasch_fit, type = "tracecontour", item = 13)
plot(mrasch_fit, type = "info")
plot(mrasch_fit, type = "SE")
plot(mrasch_fit, type = "score")
```

7.3.3 Multidimensional Graded Response Model

Both the M2PL model and the multidimensional Rasch model presented earlier are suitable for dichotomous items. It is also possible to use the MIRT framework with polytomous items. In Chapter 6, we presented the unidimensional IRT models for polytomous items with ordered and nominal response categories. Although each IRT model presented in Chapter 6 has a multidimensional form, in this section we focus, specifically, on the multidimensional graded response model (MGRM). The multidimensional forms of the other polytomous IRT models (e.g., partial credit model and rating scale model) can be estimated in a similar fashion.

In MGRM, the probability of examinee j selecting response category k in item i, θ_j can be written as

$$P(X_{ij} = k | \theta_j, a_i, \delta_{ik}) = \frac{1}{1 + \exp\left[- \sum_{m=1}^{M} = [a_{im}(\theta_{jm} - \delta_{ik})] \right]}, \quad (7.6)$$

where θ_j is a $1 \times M$ vector of latent traits for examinee j $\left(\theta_j = (\theta_{j1}, \theta_{j2}, \ldots, \theta_{jM})\right)$, a_i is a $1 \times M$ vector of slope (i.e., item discrimination) parameters for item i $\left(a_i = (a_{i1}, a_{i2}, \ldots, a_{iM})\right)$, and δ_{ik} is the category boundary location for the response category k of item i. Also, $P(X_{ij} = 0 | \theta_j, a_i, \delta_{ik}) = 1$ and $P(X_{ij} = K + 1 | \theta_j, a_i, \delta_{ik}) = 0$.

To demonstrate how to estimate MGRM, we use the **depression** data set from the **hemp** package. The **depression** data set contains the responses from 2000 examinees to 20 items on a hypothetical depression scale. Within a simple, multidimensional structure, the first 10 items measure the cognitive symptoms of depression, while the second set of 10 items measure the somatic symptoms of depression. The item names correspond to item1, item2, ..., item20 in the **depression** data set. All of the items are polytomously scored with five response categories: 0 = strongly disagree; 1 = disagree; 2 = neither agree or disagree; 3 = agree; and 4 = strongly agree).

To estimate MGRM, we again use the **mirt** function in the **mirt** package. We define two latent traits called Cognitive and Somatic using **mirt**'s model syntax and save it as mgrm_mod. Since we have a simple structure, there are no common items between the two latent traits. In the **mirt** function, we specify itemtype = "graded" to estimate MGRM based on the two-dimensional structure defined in mgrm_mod. We fit the model, extract the item parameters, and print the first few items below.

```
mgrm_mod <- 'Cognitive = 1 - 10
            Somatic = 11 - 20
            COV = Cognitive * Somatic'
```

```
mgrm_fit <- mirt(data = depression, model = mgrm_mod,
                 itemtype = "graded", SE = TRUE)
mgrm_params <- coef(mgrm_fit, simplify = TRUE)
head(mgrm_params$items)
```

```
        a1 a2      d1      d2      d3      d4
item1 1.1614  0  0.6317  0.0870 -0.9143 -1.3965
item2 1.3415  0  1.9962  1.2933  0.8925 -0.1589
item3 1.7877  0  0.2533 -0.4153 -0.9713 -1.7667
item4 1.1584  0 -0.6316 -0.9688 -1.9789 -2.3165
item5 1.1092  0  0.7715 -0.0515 -0.9521 -1.3721
item6 0.8885  0  2.3028  1.8133  1.1001  0.6449
```

In the output, the columns a1 and a2 are the slope parameters and the remaining columns are the intercept parameters (i.e., category thresholds). As we have explained before, the slope and intercept parameters are not the same as the parameters defined in Equation 7.6. To transform the estimated item parameters (i.e., slopes and intercepts) into the multidimensional item discrimination and category threshold parameters, we again use the MDISC and MDIFF functions.

```
mgrm_items <- cbind(MDISC(mgrm_fit),
                    MDIFF(mgrm_fit))
head(mgrm_items)
```

```
        MDIFF_1 MDIFF_2 MDIFF_3 MDIFF_4
item1 1.1614 -0.5439 -0.0749  0.7872  1.2024
item2 1.3415 -1.4881 -0.9641 -0.6653  0.1184
item3 1.7877 -0.1417  0.2323  0.5433  0.9883
item4 1.1584  0.5452  0.8363  1.7083  1.9997
item5 1.1092 -0.6956  0.0464  0.8584  1.2370
item6 0.8885 -2.5917 -2.0408 -1.2381 -0.7258
```

We also print the estimated variance-covariance matrix of the latent traits. The output below shows that the variances of Cognitive and Somatic are 1 and the covariance (correlation) of the two latent traits is 0.462.

```
mgrm_params$cov
```

```
          Cognitive Somatic
Cognitive 1.0000000      NA
Somatic   0.4616181       1
```

Finally, we examine the items in the **depression** data set visually. The item surface plots are slightly more complex for the multidimensional IRT models for polytomous items compared to the multidimensional IRT models for dichotomous items. In the following example, we create item surface plots for item7 and item13. Figure 7.8 shows that there are multiple surfaces in each

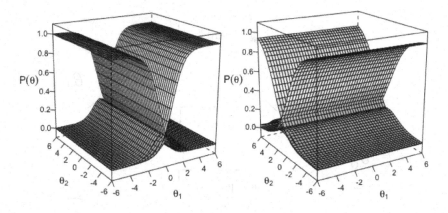

FIGURE 7.8
Item surface plots for item7 (left) and item13 (right) for the MGRM.

item surface plot because both items have multiple response categories. Each surface displays the probability of selecting a particular response category in the items.

```
itemplot(mgrm_fit, type = "trace", item = 7)
itemplot(mgrm_fit, type = "trace", item = 13)
```

We can also examine the test characteristics visually using the `plot` function. Below we present the R commands for creating various test characteristic plots. Each plot summarizes a different type of information (e.g., information, cSEM, and expected score) at the test level.

```
plot(mgrm_fit, type = "info")
plot(mgrm_fit, type = "SE")
plot(mgrm_fit, type = "score")
```

7.3.4 Bi-Factor IRT Model

The bi-factor IRT model is a special form of the MIRT modeling framework for a within-item test structure. In the bi-factor IRT models, all items are associated with a general latent trait, while each item is also associated with a secondary latent trait. The general latent trait explains the shared variability in the items, while the secondary latent traits capture the unique variability in the items. To ensure model identification, the general latent trait and the secondary latent traits are constrained to be uncorrelated in the bi-factor model. Although the bi-factor structure can be applied to a variety of MIRT models, it is particularly useful for the MIRT models where item discrimina-

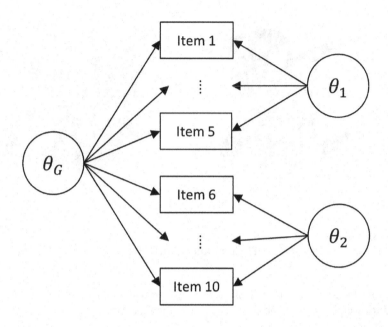

FIGURE 7.9
Bi-factor test structure.

tion parameters vary across the items (e.g., the M2PL, M3PL, and MGRM models). The bi-factor IRT model can be used for both dichotomously and polytomously scored items.

Figure 7.9 shows an example of the bi-factor structure where item 1 to item 5 are associated with the first latent trait (θ_1), item 6 to item 10 are associated with the second latent trait (θ_2), and all items are associated with a general latent trait (θ_G). The latent traits θ_1, θ_2, and θ_G are uncorrelated in the bi-factor structure.

To demonstrate how to estimate a bi-factor IRT model, we again use the `depression` data set and `bfactor` function in the **mirt** package. First, we define a vector called `dimensions` where we specify the association between each item and the two latent traits (i.e., Cognitive and Somatic). The length of this vector is the same as the number of the items in the `depression` data set. Next, we specify the test structure using `model = dimensions` in the `bfactor` function. The `bfactor` function assumes that there is an additional latent trait (i.e., a general latent trait) based on all of the items specified in the `dimensions` vector. Since the items in the `depression` data set are polytomous, the resulting model becomes the bi-factor graded response model (GRM). We fit the model, extract the estimated parameters, and print the first few parameter estimates.

```
dimensions <- rep(c(1, 2), each = 10)
bifactor_fit <- bfactor(data = depression, model = dimensions)
bifactor_params <- coef(bifactor_fit, simplify = TRUE)
head(bifactor_params$items)
```

	a1	a2	a3	d1	d2	d3	d4
item1	0.8637	0.7747	0	0.6327	0.0881	-0.9125	-1.3943
item2	0.8937	1.0200	0	2.0074	1.3015	0.8989	-0.1574
item3	1.2319	1.3070	0	0.2571	-0.4134	-0.9709	-1.7684
item4	0.8702	0.7614	0	-0.6295	-0.9662	-1.9754	-2.3128
item5	0.9003	0.6639	0	0.7747	-0.0501	-0.9529	-1.3738
item6	0.7307	0.5215	0	2.3086	1.8182	1.1037	0.6474

In the output, the first three columns (a1, a2, and a3) show the slope parameters for the general latent trait and the secondary latent traits (Cognitive and Somatic). The first slope parameter is larger than zero for all items, whereas the second and third slope parameters have zero elements for the items that are not associated with the specific latent traits. For example, item1 is associated with the general latent trait and the first secondary latent trait. Therefore, the first and second slope parameters for item1 are a1 = 0.8637 and a2 = 0.7747 while the third slope parameter is a3 = 0. The following columns (d1, d2, d3, and d4) are the intercept parameters from the bi-factor GRM.

We can also transform the estimated slope and intercept parameters into the multidimensional item discrimination and category threshold parameters using the MDISC and MDIFF functions.

```
bifactor_items <- cbind(MDISC(bifactor_fit),
                        MDIFF(bifactor_fit))
head(bifactor_items)
```

		MDIFF_1	MDIFF_2	MDIFF_3	MDIFF_4
item1	1.1602	-0.5453	-0.0760	0.7865	1.2018
item2	1.3562	-1.4801	-0.9597	-0.6628	0.1160
item3	1.7961	-0.1431	0.2302	0.5406	0.9846
item4	1.1563	0.5444	0.8356	1.7084	2.0002
item5	1.1186	-0.6926	0.0448	0.8519	1.2281
item6	0.8978	-2.5715	-2.0253	-1.2293	-0.7211

As we have explained previously, the correlations among the latent traits are constrained to be zero in the bi-factor model. We can see this constraint by printing the covariance matrix of the latent traits. In this matrix, the diagonal elements are 1 and all non-diagonal elements are zero, suggesting that the three latent traits defined by the model are orthogonal (i.e., uncorrelated).

```
bifactor_params$cov
```

```
     G S1 S2
G    1 NA NA
S1   0  1 NA
S2   0  0  1
```

7.4   Summary

In this chapter, we introduced the MIRT framework and demonstrated how to estimate and fit several MIRT models using the **mirt** package. The MIRT models are suitable for the multidimensional test structures where two or more latent traits are intended to be measured. There are several steps before applying a MIRT model to a response data set. The first step is to determine the test structure. If the test structure is unknown, an exploratory MIRT approach can be used to determine the test structure based on the response patterns. This can be considered a data-driven approach. If the relationship between the items and the latent traits is already known, then a confirmatory MIRT approach should be used. The test structure can exhibit either within-item multidimensionality or between-item multidimensionality. The second step is to determine whether there is a compensatory relationship between the latent traits. The compensatory MIRT models allow the strength in one latent trait to compensate for the weakness in another latent trait. The non-compensatory MIRT models do not allow any compensation among the latent traits. The third step is to select a MIRT model. The unidimensional IRT models described in Chapters 5 and 6 can be estimated as a MIRT model. This means that the MIRT models can be used with both dichotomous and polytomous items. Although the **mirt** package is capable of estimating a variety of MIRT models, we have only demonstrated the common MIRT models used in educational and psychological measurement. We recommend our readers check out the **mirt** package manual (Chalmers, 2012) for more details about the capabilities of **mirt**.

8

Explanatory Item Response Theory

8.1 Chapter Overview

In Chapters 5 trough 7, we described unidimensional and multidimensional IRT models for dichotomously and polytomously scored items. These IRT models, regardless of the number of response categories, provide direct information regarding examinees' latent trait levels and information about each item concerning the difficulty, discrimination, and guessing, depending on the selected IRT model. Despite being informative about individual items and examinees, traditional IRT models are not capable of explaining common variability across items or examinees based on the design or the theory behind a measurement instrument. De Boeck and Wilson (2004) introduced explanatory IRT modeling (EIRM) that can be used for measuring common variability in item clusters, examinee groups, or the interactions between item clusters and examinee groups. Unlike traditional IRT models, EIRM aims to explain the commonality of item responses using explanatory variables (also known as covariates) related to items (e.g., content, cognitive complexity, and item format) and examinees (e.g., gender, ethnicity, and grade level). Chapter 8 provides a summary of various explanatory IRT models and demonstrates how to estimate these models using the **lme4** package (Bates et al., 2015) in R. Explanatory IRT models can also be estimated with the **mirt** (Chalmers, 2012), **flirt** (Jeon et al., 2014), and **eRm** (Mair & Hatzinger, 2007a) packages.

8.2 Explanatory Item Response Modeling

Explanatory IRT modeling (EIRM) can utilize IRT for both measurement and explanatory purposes (De Boeck & Wilson, 2004). The main advantage of EIRM over traditional IRT models is its flexibility to analyze items and examinees, while simultaneously decomposing common variability across items and examinees (Briggs, 2008). Under the EIRM framework, traditional IRT models are considered to be a subset of models that belong to a larger class of models — generalized linear mixed models (GLMMs). GLMMs can function as explanatory IRT models when the model includes explanatory variables as predictors of item responses. These explanatory variables can explain common variability in item clusters, examinee groups, or the interactions between items and examinees (De Boeck & Wilson, 2004).

The EIRM approach defines persons as clusters, items as the repeated observations, and item responses as the dependent variable within a multilevel structure. Using explanatory covariates, the dependent variable (i.e., item responses) can be predicted in a more parsimonious way. For example, assume that we have a math test that consists of two types of items: algebra items with mathematical expressions, notations, and terms, and word problems that describe the math problem verbally within a scenario. When we estimate the item difficulty parameters, we can use item type (i.e., algebra items and word problems in a math test) as a covariate to explain the variability in the item difficulty with regard to item type (instead of estimating a unique item difficulty parameter for each item). This would allow us to examine whether algebra items are likely to be more difficult or easier than word problems. Then we can explain the overall impact of item type on the items.

In the following sections, we demonstrate three applications of EIRM; namely, the linear logistic test model, the latent regression model, and the interaction model. For brevity, we focus only on dichotomous item responses, although it is possible to use EIRM with polytomous item responses. It should be noted that the three applications mentioned above are a limited subset of what is possible within the EIRM framework. A more comprehensive discussion of the EIRM framework can be found in other resources (e.g., De Boeck & Wilson, 2004).

8.2.1   Data Structure

In this chapter, we use the **lme4** (Bates et al., 2015) package for estimating explanatory IRT models. This is the same package that we used when we discussed generalizability theory in Chapter 3. So far, we have demonstrated the use of the **mirt** package for estimating various IRT models. The **mirt** package requires item response data to be organized in a wide or tidy (Wickham, 2017) format where there is one row per examinee and one column per item. Unlike the **mirt** package, the **lme4** package requires the data to be organized in a long format where there is one row for each unique combination of an examinee and an item. For example, the long format for a 10-item test with 500 examinees would have a single column with 5000 rows (10 items x 500 examinees) of item responses. Additional identifying variables (e.g., item and examinee identifiers) are also necessary for the estimation process in the **lme4** package.

In the following example, we use the `multiplechoice` data set from the **hemp** package to demonstrate how to transform a data set from a wide format to a long format. The `multiplechoice` data set consists of 27 columns of responses to 27 items labelled item1 through item27. There are 496 rows, indicating there are 496 examinees in the data set. We use the `reshape` function for this transformation. The `reshape` function requires several arguments:

- the name of the data set we wish to reshape, `multiplechoice`;

- the columns of the original data set that contain the responses, in this case columns 1 through 27, `varying = 1:27`;

- the target direction for our reshaped data, `direction = "long"`;

- the name of the new time variable; in this case the time variable is actually called item because we have repeated measurements on items not time, `timevar = "item"`; and

- the name for the new (or existing) identifier variable, `idvar = "id"`; and finally the name of the new variable that contains the item responses, `v.names = "response"`.

We save the transformed data set as a new data set called `mc_long` where each examinee has 27 rows (i.e., one row per item in the `multiplechoice` data set).

```
mc_long <- reshape(multiplechoice,
                   varying = 1:27,
                   direction = "long",
                   timevar = "item",
                   idvar = "id",
                   v.names = "response")
```

The `mc_long` data set is sorted by "item" by default. We would like to sort the data by the examinee identifier (i.e., id) first and then the item identifier (i.e., item). We can do this using the `order` function. We print the first six rows of the long-format data set using the `head` function.

```
mc_long <- mc_long[order(mc_long$id, mc_long$item), ]
head(mc_long)
```

```
    item response id
1.1    1        4  1
1.2    2        1  1
1.3    3        2  1
1.4    4        3  1
1.5    5        4  1
1.6    6        3  1
```

8.2.2   Rasch Model as a GLMM

In Chapter 5, we presented the Rasch model as a special case of the 1PL IRT model. The Rasch model predicts the probability of responding to an item correctly given item difficulty and examinees' latent trait levels. In the Rasch model, the item discrimination parameter is assumed to be 1 for all items and there is no model estimated lower asymptote (i.e., no guessing expected). The Rasch model can be defined as a GLMM where the dependent variable (the dichotomous item responses) is predicted by some fixed effects (items) along with random effects (examinees' latent trait levels).

For a GLMM, item responses are denoted as $Y_{ij} = 0$ or $Y_{ij} = 1$, where $j = 1, \ldots, J$ is the index for examinees and $i = 1, \ldots, K$ is the index for items. Y_{ij} has a Bernoulli distribution with mean, π_{ij}. A logit link function[1] is typically applied to place π_{ij} onto a continuous scale between $-\infty$ and $+\infty$ as:

$$\eta_{ij} = \log\left(\frac{\pi_{ij}}{1 - \pi_{ij}}\right). \tag{8.1}$$

Following the notation in De Boeck and Wilson (2004), the GLMM formulation of the Rasch model can be written as:

$$\eta_{ij} = \theta_j + \sum_{k=1}^{K} \beta_i X_{ik}, \tag{8.2}$$

where θ_j is the latent trait level for examinee j, β_i refers to the item easiness for item i ($i = 1, 2, \ldots, K$) as opposed to item difficulty in the traditional IRT models, and $X_{ik} = 1$ if $i = k$ and 0 otherwise. In Equation 8.2, θ_j is the random intercept with a normal distribution, i.e. $\theta_j \sim N(0, \sigma^2)$ and X_{ik} is the item indicator serving as the fixed intercept in the model. Equation 8.2 can be written more compactly as:

$$\eta_{ij} = \theta_j + \beta_i, \tag{8.3}$$

where β_i is the sum of the product of β_i and X_{ik}. In Equation 8.3, β_i can be multiplied by -1 to reverse the sign of the item easiness parameter, which would change its interpretation from item easiness to item difficulty.

To demonstrate how to estimate the explanatory IRT models, we use the **eirm** data set from the **hemp** package. Before we begin the analysis, we first activate the **hemp** and **lme4** packages using the `library` command.

---

[1] A probit link function can also be used.

```
library("hemp")
library("lme4")
```

The `eirm` data set contains the responses of 1000 examinees to a hypothetical quiz with ten items. The items were scored dichotomously (1 = correct; 0 = incorrect). The `eirm` data set is already in long format, and thus, there is no need for a data transformation. In addition to the three columns identifying examinees (person), items (item), and responses to the items (response), there are two explanatory variables: gender (F = female; M = male) and itemtype (Nonvisual or Visual), which pertains to whether the items included any visual element (e.g., a graph, a chart, or an image). Both gender and itemtype are character variables in the data set. We print the first ten rows of the `eirm` data set with the `head` function.

```
head(eirm, 10)
```

	person	item	response	gender	itemtype
1	1	1	0	F	Nonvisual
2	1	2	0	F	Nonvisual
3	1	3	0	F	Nonvisual
4	1	4	0	F	Nonvisual
5	1	5	0	F	Nonvisual
6	1	6	0	F	Visual
7	1	7	0	F	Visual
8	1	8	0	F	Visual
9	1	9	0	F	Visual
10	1	10	0	F	Visual

Next, we use the `table` function to create a two-way table of examinee responses to items by item type.

```
table(eirm$item, eirm$itemtype)
```

	Nonvisual	Visual
1	1000	0
2	1000	0
3	1000	0
4	1000	0
5	1000	0
6	0	1000
7	0	1000
8	0	1000
9	0	1000
10	0	1000

The output above shows that the first five items in the `eirm` data set are nonvisual items, while the last five items involve a visual element. All examinees responded to all items on the quiz (i.e., items and examinees are fully crossed; see Chapter 3 for a refresher of a crossed design). The examinees were also balanced with regards to gender.

We begin the IRT analysis of the `eirm` data set with the Rasch model. When estimating the Rasch model, we ignore the impact of the explanatory variables and focus on the individual items and examinees. To estimate the Rasch model, we use the `glmer` function in the **lme4** package. The `glmer` function is capable of estimating generalized linear mixed models with fixed and random effects. Unlike in Chapter 3, we begin by specifying the optimizer and the maximum number of function evaluations before using the `glmer` function. Although the following settings are sufficient for estimating most explanatory IRT models, we encourage readers to look over the **lme4** package manual for more details on the control options.

```
control <- glmerControl(optimizer = "bobyqa",
                        optCtrl = list(maxfun = 100000))
```

Next, we use the `glmer` function to estimate the Rasch model and save the results as `rasch_mod`. Below in our call to the `glmer` function, `response` ~ defines response as the dependent variable in the model; `-1 + item` defines item as a fixed effect to predict response, the -1 eliminates the intercept in the model; the `(1 | person)` defines examinees as the random effects; `family = binomial` specifies a binomial distribution with a logit link function; and `control = control` activates the control statement we defined above.

```
rasch_mod <- glmer(response ~ -1 + item + (1 | person),
                   family = binomial, data = eirm,
                   control = control)
```

Next we save the estimated parameters in a data frame called `rasch_params`. The `fixef` function retrieves the estimated fixed effects (the item parameters in the Rasch model) from `rasch_mod`.

```
rasch_params <- data.frame(easiness = fixef(rasch_mod))
rasch_params
```

```
          easiness
item1    1.7265635
item2    0.8168178
item3    0.4346300
item4    0.4157493
item5    0.2385410
item6   -0.2482347
item7   -0.7259675
item8   -0.9741741
item9   -0.9741724
item10  -1.3409822
```

As we explained earlier, the parameters obtained from the `glmer` function represent item easiness rather than item difficulty due to the positive parameterization of item difficulty in the GLMM (see Equation 8.3). The output shows that item 1 is the easiest item and item 10 is the most difficult item in the `eirm` data set. In addition to the item difficulty parameters, we can also extract the random effects (latent trait estimates) from `rasch_mod`. We use the `coef` function to retrieve only the first column of the estimated coefficients for person because the remaining columns are the item parameters. We save the latent trait estimates as `rasch_ability`.

```
rasch_ability <- coef(rasch_mod)$person[, 1]
```

Lastly, we use both item parameters and latent trait estimates in the `itemperson_map` function from the **hemp** package to draw an item-person map for the Rasch model. We first save the item parameters into a new data set called `difficulty` and rename the columns as 1 through 10 (referring to item 1 to item 10). Then, we use both `difficulty` and `rasch_ability` to draw an item-person map. Figure 8.1 shows the distribution of the item easiness parameters relative to the examinees' latent trait levels. This plot is particularly useful for reviewing whether the difficulty range of a test is adequate given the range of the latent trait levels among the examinees.

```
difficulty <- rasch_params$easiness
names(difficulty) <- 1:10
itemperson_map(difficulty, rasch_ability, n = 50)
```

De Boeck and Wilson (2004) described the Rasch model as a doubly-descriptive model because each item has a unique difficulty parameter and each examinee has a unique latent trait estimate. When predictors explaining the individual effects of the items and of the examinees are incorporated into the Rasch model, the resulting model becomes an explanatory IRT model, which is often a more parsimonious form of the Rasch model (i.e., fewer parameters to explain the items). Next, we illustrate how to transform the traditional Rasch model into an explanatory IRT model by including item-related and examinee-related predictors.

8.2.3 Linear Logistic Test Model

The *linear logistic test model* (LLTM; Fischer, 1973) incorporates item properties as predictors to explain the variation among items regarding their effect on the probability of success (i.e., η_{ij}). Unlike unique item difficulty parameters in the Rasch model, LLTM involves fewer difficulty parameters based on the item properties. The mathematical formulation of LLTM is:

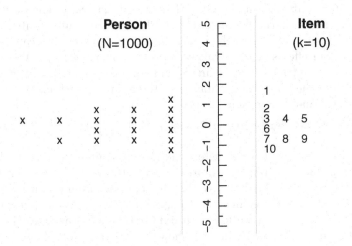

FIGURE 8.1
Item-person map for the Rasch model.

$$\eta_{ij} = \theta_j + \sum_{m=1}^{M} \beta_m X_{im}, \tag{8.4}$$

where η_{ij} is examinee j's probability of success on item i (placed on a logit scale), θ_j is the latent trait level for examinee j, X_{im} is the value of item i on item property m ($m = 1, 2, \ldots, M$), and β_m is the regression weight of item property m. In LLTM, multiple item property variables and their interactions can be used together as predictors. Because the number of item property variables (M) is expected to be less than the number of items on the test (K), the prediction is likely to be less precise. That is, $\sum_{m=1}^{M} \beta_m X_{im}$ in Equation 8.4 will not be the same as β_i in Equation 8.3. However, depending on the quality of the item properties, a highly precise prediction can still be achieved with fewer difficulty parameters.

Alternatively, a residual term can be included in LLTM. The resulting model is often called the LLTM with error. This model requires items to be treated as random effects in the estimation process. The LLTM with error allows for an imperfect prediction as opposed to a regression-based, perfect prediction with the regular LLTM (De Boeck, 2008; Doran, Bates, Bliese, & Dowling, 2007). The mathematical formulation of the LLTM with error is:

$$\eta_{ij} = \theta_j + \sum_{m=1}^{M} \beta_m X_{im} + \epsilon_i, \tag{8.5}$$

where ϵ_i is a normally-distributed residual term, $\epsilon_i \sim N(0, \sigma_\epsilon^2)$, and the remaining elements are the same as those from Equation 8.4. The LLTM with error assumes homoscedasticity for the residual variance (De Boeck et al., 2011).

To demonstrate how to estimate LLTM and the LLTM with error, we use the "itemtype" variable in the **eirm** data set to predict the probability of successfully responding to the items. Instead of estimating individual item difficulty parameters, we will estimate a single difficulty parameter that explains item difficulty depending on whether the items include a visual element. That is, we use itemtype in the **eirm** data set as an item-related predictor in LLTM.

To estimate LLTM with the **lme4** package, we use the **glmer** function again. The code is very similar to the one for the Rasch model with the exception that itemtype is used instead of item. With this modification, we estimate two item difficulty parameters (one for nonvisual and one for visual items). Because there are only two item difficulty parameters (as opposed to ten parameters in the Rasch model), the estimation of LLTM is slightly faster. We print the results using the **summary** function.

```
lltm_mod <- glmer(response ~ -1 + itemtype + (1 | person),
                  family = binomial, data = eirm,
                  control = control)
summary(lltm_mod)


Random effects:
 Groups Name          Variance Std.Dev.
 person (Intercept) 0.548     0.7403
Number of obs: 10000, groups:  person, 1000


Fixed effects:
                   Estimate Std. Error z value Pr(>|z|)
itemtypeNonvisual   0.67779    0.03949   17.16   <2e-16 ***
itemtypeVisual     -0.82424    0.04014  -20.53   <2e-16 ***
---
Signif. codes:
0 '***' 0.001 '**' 0.01 '*' 0.05 '.' 0.1 ' ' 1
```

We printed just some of the output from the **glmer** function above, specifically the random effects and fixed effects sections. LLTM provides an estimate of the variance for person as $\sigma_{\theta_j}^2 = 0.548$, which indicates the variation among examinees in terms of their latent trait levels. The estimated difficulty parameters for the visual and nonvisual items are 0.677 and -0.824, respectively.

As we mentioned earlier, the item parameters from the `glmer` function represent item easiness. In other words, higher values indicate easier items. The results show that the nonvisual items are easier than the visual items in the `eirm` data set. In the same table, we see the standard errors, the z statistic, and corresponding p-values for the estimated parameters. These significance tests indicate whether the estimated parameters are significantly different from zero. In our example, both difficulty parameters are significantly different from zero at the alpha level of $\alpha = .05$.

Next, we estimate the LLTM with error. For this model, we add item into the model as random effects. The LLTM with error allows item difficulty parameters and latent trait estimates to vary as random effects while controlling for the impact of item type as an explanatory variable. We save the estimation results as `lltme_mod` and view the results using the `summary` function.

```
lltme_mod <- glmer(response ~ -1 + itemtype + (1 | person) +
                   (1 | item), family = binomial, data = eirm,
                   control = control)
summary(lltme_mod)
```

```
Random effects:
 Groups Name          Variance Std.Dev.
 person (Intercept) 0.6124    0.7825
 item   (Intercept) 0.1983    0.4453
Number of obs: 10000, groups:  person, 1000; item, 10

Fixed effects:
                  Estimate Std. Error z value Pr(>|z|)
itemtypeNonvisual   0.7218     0.2033   3.550 0.000385 ***
itemtypeVisual     -0.8495     0.2033  -4.178 2.95e-05 ***
---
Signif. codes:
0 '***' 0.001 '**' 0.01 '*' 0.05 '.' 0.1 ' ' 1
```

Again we show some of the output returned from the `glmer` function. With the addition of item, the random effects section shows the variances of both the latent trait and the items. Comparing this output with the LLTM, the variance of the latent trait in the LLTM with error is larger ($\sigma^2_{\theta_j} = 0.612$) because the LLTM explains the variation between the items based on only itemtype, assuming that the residual term is zero. The estimate of the item variance is 0.198. The fixed effects section shows the estimated effects of visual and nonvisual items. The estimated effects from the LLTM with error are slightly larger than those from the LLTM.

We can use the `anova` function to compare the Rasch model and the two explanatory IRT models that we have estimated so far. The `anova` function provides several goodness-of-fit statistics (e.g., log-likelihood, AIC, and BIC).

Although the **anova** function also provides a likelihood ratio test, this test is not appropriate here because the Rasch model and the explanatory IRT models are not nested within each other. First, we compare the Rasch model to the LLTM.

```
anova(rasch_mod, lltm_mod)
```

	Df	AIC	BIC	logLik	deviance
lltm_mod	3	12496	12518	-6245.0	12490
rasch_mod	11	12142	12221	-6059.8	12120

The output shows that the Rasch model has the smaller log-likelihood, AIC, and BIC values, suggesting that it fits the data better than the LLTM. This result is not surprising, because the Rasch model provides a unique difficulty parameter for each item and a variance estimate for the latent trait. Despite indicating a poorer model-fit, the LLTM is a more parsimonious model given that it only has three parameters (2 item parameters and a variance for the latent trait).

The second comparison is between the Rasch model and the LLTM with error. The LLTM with error includes a unique item difficulty parameter for each item and two additional parameters for itemtype (i.e., visual and nonvisual). Therefore, the LLTM with error is expected to provide a better model fit than LLTM.

```
anova(rasch_mod, lltme_mod)
```

	Df	AIC	BIC	logLik	deviance
lltme_mod	4	12178	12207	-6085.0	12170
rasch_mod	11	12142	12221	-6059.8	12120

The output from the second comparison shows that the Rasch model has a better fit than the LLTM with error based on the AIC and log-likelihood values. However, the addition of the residual term into LLTM resulted in a better goodness-of-fit, which is reflected in the lowest BIC value among the three models. The LLTM with error appears to be a better option for examining the effects of the item property variables because it takes the variability of the items into account, which leads to more accurate parameter estimates.

8.2.4 Latent Regression Rasch Model

The latent regression Rasch model (Zwinderman, 1991) incorporates examinee-related variables into the Rasch model to explain differences between examinees with regard to the latent trait being measured. In a GLMM, examinee-related predictors are typically used as fixed effects, although their effects can also be estimated as random effects. The mathematical formulation of the latent regression Rasch model is as follows:

$$\eta_{ij} = \left(\sum_{p=1}^{P} \vartheta_p Z_{jp} + \theta_j \right) + \beta_i, \tag{8.6}$$

where η_{ij} is examinee j's probability of answering item i correctly (on the logit scale), Z_{jp} is the value of examinee j on examinee property p ($p = 1, 2, \ldots, P$), ϑ_p is the regression weight of examinee property p, θ_j is the latent trait level for examinee j after the effect of the examinee properties is accounted for, and β_i is the difficulty parameter for item i (i.e., item easiness in GLMM). Multiple examinee property variables (e.g., gender, ethnicity, and SES) as well as their interactions can be used together as predictors in the latent regression Rasch model.

To demonstrate how to estimate the latent regression Rasch model, we use the gender variable in the `eirm` data set to predict the probability of responding correctly to the items. The two fixed effects in the model are item, to estimate the item difficulty parameters, and gender, to estimate the effect of examinees' gender. The gender variable is a character variable with two values: F for female examinees and M for male examinees. For the gender variable, females are the reference group.[2]

```
lrrm_mod <- glmer(response ~ -1 + item + gender + (1 | person),
                  family = binomial, data = eirm,
                  control = control)
summary(lrrm_mod)
```

```
Random effects:
 Groups Name             Variance Std.Dev.
 person (Intercept) 0.4626    0.6802
Number of obs: 10000, groups:  person, 1000

Fixed effects:
         Estimate Std. Error z value Pr(>|z|)
item1     1.32578    0.09399   14.106  < 2e-16 ***
item2     0.41592    0.08043    5.171 2.33e-07 ***
item3     0.03392    0.07815    0.434   0.6643
item4     0.01505    0.07808    0.193   0.8472
item5    -0.16203    0.07765   -2.087   0.0369 *
item6    -0.64844    0.07830   -8.281  < 2e-16 ***
item7    -1.12592    0.08149  -13.817  < 2e-16 ***
item8    -1.37408    0.08417  -16.325  < 2e-16 ***
item9    -1.37408    0.08417  -16.324  < 2e-16 ***
item10   -1.74095    0.08955  -19.441  < 2e-16 ***
genderM   0.80083    0.06325   12.662  < 2e-16 ***
```

---

[2]By default, R chooses the first value in alphabetical order as the reference category.

The estimated effect of gender was 0.801, which is statistically significant at an alpha level of $\alpha = .001$. The positive effect for gender suggests that the male examinees are more likely to answer the items correctly compared to the female examinees in the `eirm` data set.

In addition to the estimated fixed effect for gender, the random effects show that the estimated variance of the latent trait estimates (0.463) is less than the latent trait variance from the Rasch model (0.622). This finding suggests that gender explains some part of the latent trait variation among the examinees. The latent regression Rasch model assumes that the effects of examinee property variables are fixed, and thus they do not include error. This assumption can be partially relaxed by adding a random-effect component for the examinee property variables. In the following example, we include gender in the random effects section using `(gender | person)`. This change allows the effect of gender to vary across the examinees in addition to the fixed effects for gender.

```
lrrme_mod <- glmer(response ~ -1 + item + gender +
                   (gender | person), family = binomial,
                   data = eirm, control = control)
summary(lrrme_mod)
```

```
Random effects:
 Groups Name          Variance Std.Dev. Corr
 person (Intercept) 0.4696   0.6853
        genderM       0.5107   0.7147   -0.54
Number of obs: 10000, groups:  person, 1000


Fixed effects:
         Estimate Std. Error z value Pr(>|z|)
item1     1.32583    0.09408  14.092  < 2e-16 ***
item2     0.41553    0.08057   5.157 2.51e-07 ***
item3     0.03341    0.07832   0.427   0.6697
item4     0.01454    0.07826   0.186   0.8526
item5    -0.16258    0.07784  -2.089   0.0367 *
item6    -0.64899    0.07849  -8.269  < 2e-16 ***
item7    -1.12639    0.08163 -13.798  < 2e-16 ***
item8    -1.37447    0.08429 -16.306  < 2e-16 ***
item9    -1.37447    0.08429 -16.306  < 2e-16 ***
item10   -1.74118    0.08963 -19.427  < 2e-16 ***
genderM   0.80103    0.06327  12.661  < 2e-16 ***
```

The output above also shows that the variance estimate for gender is 0.511 and the correlation between the latent trait estimates and the random effects for gender is $r = -.54$. Next, we use the `anova` function to determine whether the fixed and random effects of gender made significant contributions to the

Rasch model (i.e., the baseline model). The first comparison evaluates the impact of gender as a fixed-effect variable, while the second comparison evaluates the impact of the varying effect of gender as a random-effect variable.

```
anova(rasch_mod, lrrm_mod)
anova(lrrm_mod, lrrme_mod)
```

	Df	AIC	BIC	logLik	deviance
rasch_mod	11	12142	12221	-6059.8	12120
lrrm_mod	12	11990	12077	-5983.1	11966

	Df	AIC	BIC	logLik	deviance
lrrm_mod	12	11990	12077	-5983.1	11966
lrrme_mod	14	11994	12095	-5983.1	11966

The output shows that using gender as a fixed-effect variable in the latent regression Rasch model improved the fit statistics relative to the Rasch model. This is an expected finding because of the statistically significant effect for gender in the latent regression Rasch model. The second comparison shows that using gender as a varying random-effect variable in lrrme_mod did not improve the model fit. With two additional parameters (the variance estimate for gender and the correlation between gender and the latent trait estimates), lrrme_mod did not have better fit statistics compared to the latent regression Rasch model. This finding suggests that the effect of gender does not seem to vary significantly across the examinees.

As previously mentioned, there are multiple ways of including examinee property variables in the latent regression Rasch model. For example, the heteroscedasticity of the unexplained latent trait variance can be modeled with only varying effects for the examinee property variable, add (-1 + gender | person). Alternatively, a multilevel Rasch model can be estimated using the examinee property variable as a third level (1 | gender) in addition to the items (level 1) and the examinees (level 2). A more detailed description of various latent regression Rasch models can be found in De Boeck et al. (2011).

8.2.5   Interaction Models

As we described at the beginning of this chapter, it is possible to include both item property variables and examinee property variables within the same explanatory IRT model. For example, one can combine LLTM in Equation 8.4 and the latent regression Rasch model in Equation 8.6. This yields the *latent regression LLTM*, which has both item and examinee-related explanatory variables. The latent regression LLTM can be formulated as:

$$\eta_{ij} = \left(\sum_{p=1}^{P} \vartheta_p Z_{jp} + \theta_j \right) + \sum_{m=1}^{M} \beta_m X_{im}. \tag{8.7}$$

A residual term (ϵ_i) can be included in Equation 8.7 to estimate the varying effects of items (i.e., unique item difficulty parameters). In addition, the product of an item property variable (or the items) and an examinee property variable can be included as an interaction term in the model. This interaction term can be a fixed-effect or random-effect variable. Equation 8.8 shows an interaction model with the product of item property variables and examinee property variables as fixed-effect variables:

$$\eta_{ij} = \left(\sum_{p=1}^{P} \vartheta_p Z_{jp} + \theta_j \right) + \sum_{m=1}^{M} \beta_m X_{im} + \sum_{h=1}^{H} \omega_h W_{(ij)h}, \tag{8.8}$$

where ω_h is the regression weight for the interaction term, $W_{(ij)h}$ is the interaction variable, and the remaining elements are the same as those from Equation 8.7. For example, the product of itemtype with two levels (1 = Nonvisual, 0 = Visual) and gender with two levels (0 = Female, 1 = Male) yields $W_{(ij)h} = 1$, if both itemtype = 1 and gender = 1, and $W_{(ij)h} = 0$ otherwise.

In the following section, we present two examples of the interaction models using the `eirm` data set. First, we estimate an interaction model with itemtype, gender, and their interaction. This model demonstrates the differential effects of itemtype depending on gender. In the second example, we estimate a latent regression Rasch model with gender and its interaction with item. This model examines whether the response probabilities of examinees with the similar latent trait levels differ depending on the gender group they belong to. Individual items showing significant interactions with gender are then flagged for differential item functioning (DIF). We discuss the details of DIF and measurement invariance in Chapter 11.

The first interaction model presented below is a combination of the LLTM with error and the latent regression Rasch model. In this model, we use `itemtype * gender` to estimate the main effects of itemtype and gender as well as an additional fixed effect for their interaction. We save the results from this interaction model as `int_mod`.

```
int_mod <- glmer(response ~ -1 + itemtype * gender +
                 (1 | person) + (1 | item), family = binomial,
                 data = eirm, control = control)
summary(int_mod)
```

```
Random effects:
 Groups Name            Variance Std.Dev.
 person (Intercept) 0.4554    0.6748
 item   (Intercept) 0.2000    0.4472
Number of obs: 10000, groups:  person, 1000; item, 10

Fixed effects:
                        Estimate Std. Error z value Pr(>|z|)
itemtypeNonvisual        0.27440    0.20671   1.327   0.1844
itemtypeVisual          -1.18848    0.20789  -5.717 1.09e-08 ***
genderM                  0.90748    0.07778  11.667  < 2e-16 ***
itemtypeVisual:genderM  -0.21751    0.09165  -2.373   0.0176 *
```

The results from the interaction model are similar to those from the previous models regarding the main effects of the itemtype and gender variables. The visual items are more difficult than the nonvisual items. Also, the male examinees are more likely to respond to the items correctly than the female examinees. The final row labeled as `itemtypeVisual:genderM` indicates the interaction term in the model. The estimated regression weight for the interaction term is -0.218, which is statistically significant at an $\alpha = .05$ level of significance. From the sign of this regression weight, we can conclude that the male examinees are less likely to answer the visual items correctly compared to the female examinees, regardless of their latent trait levels.

The second example focuses on the interaction between the individual items and gender. As mentioned earlier, a significant interaction term between a particular item and gender indicates uniform DIF. Uniform DIF means that examinees from a particular group (defined by a particular examinee property variable, such as gender) consistently perform better than the examinees in another group. Using the `eirm` data set, we examinee whether the items exhibit uniform DIF. We use `item * gender` to estimate the main effects for item and gender as well as their interactions. We save the model results as `dif_mod`.

```
dif_mod <- glmer(response ~ -1 + item * gender +
                 (1 | person), family = binomial,
                 data = eirm, control = control)
summary(dif_mod)
```

```
Fixed effects:
        Estimate Std. Error z value Pr(>|z|)
item1   1.316780   0.115085  11.442  < 2e-16 ***
item2   0.276162   0.099790   2.767  0.00565 **
item3   0.008137   0.099069   0.082  0.93454
item4   0.034803   0.099084   0.351  0.72540
item5  -0.241584   0.099508  -2.428  0.01519 *
item6  -0.686260   0.103050  -6.659 2.75e-11 ***
```

```
item7           -1.280308  0.113640 -11.266  < 2e-16 ***
item8           -1.198432  0.111739 -10.725  < 2e-16 ***
item9           -1.316332  0.114505 -11.496  < 2e-16 ***
item10          -1.415466  0.117068 -12.091  < 2e-16 ***
genderM          0.829131  0.183432   4.520 6.18e-06 ***
item2:genderM    0.300560  0.230788   1.302  0.19281
item3:genderM    0.029139  0.225234   0.129  0.89706
item4:genderM   -0.068110  0.224730  -0.303  0.76183
item5:genderM    0.139082  0.224591   0.619  0.53574
item6:genderM    0.045159  0.224506   0.201  0.84058
item7:genderM    0.246283  0.229620   1.073  0.28346
item8:genderM   -0.346861  0.230583  -1.504  0.13251
item9:genderM   -0.132313  0.231356  -0.572  0.56739
item10:genderM  -0.614679  0.237048  -2.593  0.00951 **
```

The first part of the output shows the item difficulty parameters and the fixed effect for gender. We focus on the second part of the output showing the interactions between the individual items and gender. The only significant interaction, at an $\alpha = .01$ level of significance, is `item10:genderM`. From the sign of this interaction, we can conclude that the male examinees are less likely to answer item 10 correctly than the female examinees. That is, compared to the female examinees, the male examinees find item 10 more difficult.

Instead of an omnibus test of all interactions together, we could also test each item separately by including a single interaction term between a particular item and gender. For example, we can create a dummy variable (called dif10) for item 10 where dif10 is equal to 1 if item (i.e., item identifier) is 10 and gender is M, and 0 otherwise. Instead of `item * gender`, we need to use `item + gender + dif10` to estimate the item difficulty parameters, the gender effect, and the interaction effect for item 10. Then, we can compare the likelihood of this model with the likelihood of the Rasch model. The likelihood ratio test from the **anova** function is suitable for this comparison because the Rasch model is nested within the interaction model.

```
dif10 <- with(eirm,
            ifelse(gender == "M" & item == "10", 1, 0))
dif10_mod <- glmer(response ~ -1 + item + gender +
                  dif10 + (1 | person), family = binomial,
                  data = eirm, control = control)
anova(rasch_mod, dif10_mod)
```

```
           Df  AIC   BIC   logLik deviance  Chisq Df Pr(>Chisq)
rasch_mod  11 12142 12221 -6059.8    12120
dif10_mod  13 11977 12071 -5975.6    11951 168.57  2  < 2.2e-16
```

A significant χ^2 statistic from the likelihood ratio test indicates that the larger model with more parameters fits the data better than the smaller model with fewer parameters. In the output above, we see that the difference between `dif10_mod` (more complex model) and `rasch_mod` (baseline model) is statistically significant, $\chi^2(2) = 168.57, p < .001$, indicating a better model fit for `dif10_mod`. This finding also suggests that the estimated DIF parameter for item 10 is statistically significant because this is the only difference between the two models. This process can be repeated for each item to detect the items exhibiting uniform DIF.

8.3   Summary

In this chapter, we introduced the EIRM framework and demonstrated how to estimate several explanatory IRT models using the **lme4** package. The GLMM framework provides a great deal of flexibility for examining the effects of item property variables, examinee property variables, and their interactions in the context of EIRM. The models we have presented in this chapter illustrate the main differences between the Rasch model and explanatory IRT models. Using the `glmer` function in the **lme4** package, it is possible to estimate the effects of the explanatory variables as either fixed or random effects. Different combinations of the fixed and random effects can result in a large variety of explanatory IRT models. A more comprehensive discussion of the EIRM framework and the `glmer` function can be found in De Boeck and Wilson (2004) and De Boeck et al. (2011).

Although the `glmer` function is a great tool for estimating a variety of explanatory IRT models, it can only estimate the Rasch model and its explanatory variants. The `glmer` function cannot be used for estimating the other IRT models presented in Chapters 5 and 6 (e.g., 2PL model, 3PL model, and GRM). Also, in comparison with other R packages, the `glmer` function often requires a longer estimation time for complex models with large numbers of fixed and random effects. Therefore, we recommend the **mirt** (Chalmers, 2012) and **flirt** (Jeon et al., 2014) packages if explanatory forms of the IRT models other than the Rasch model are needed and there are large numbers of items and examinees in the response data set.

9

Visualizing Data and Measurement Models

9.1 Chapter Overview

In this chapter, we focus on the graphical representations of data and measurement models using R. We begin by showing how to create diagnostic plots that could be used for factor analysis and item response theory (IRT) models. The plots presented in this chapter allow us to assess the plausibility of our modeling assumptions and to look for observations that might be outliers or influential. Next, we show how to create path diagrams that could be used for publications or presentations with the **semPlot** package (Epskamp & Stuber, 2017). Finally, we present two applications of interactive data visualization with the **shiny** package (Chang, Cheng, Allaire, Xie, & McPherson, 2017). The first application can help further one's understanding of a diagnostic plot presented in this chapter, and the second application shows how to explore the impact of different parameters on the shape of the item characteristic curve in unidimensional IRT models.

9.2 Introduction

Graphical data analysis is an important component of statistical analysis. Exploration of the data, especially prior to the modeling step, is essential for identifying, confirming, and describing (e.g., linear or nonlinear) relationships between variables and for identifying potential influential observations or outliers. Once we fit a statistical or measurement model to the data, we use graphical analysis to assess the appropriateness of our modeling assumptions, the fit of the model to the data, the overall adequacy of the model at describing the relationships within the data, and to confirm whether influential observations and outliers have any impact on the model results.

Graphs represent an excellent medium for communicating and understanding model findings either formally (e.g., in manuscripts and presentations) or informally (e.g., to colleagues or for ourselves). Graphs are typically less cognitively demanding than tables and can be used to summarize and explain complex relationships that would be difficult to demonstrate in tables or that would take up substantial space in our publications. Graphs are also a great tool for learning about the measurement models that we have explained throughout this book. For example, we could examine how an item characteristic curve (ICC) from a three-parameter logistic (3PL) IRT model changes as a function of different values of the item difficulty, item discrimination, and guessing parameters. To demonstrate this, we would plot the ICC based on user specified values for the model parameters. This would allow us to understand how difficulty, discrimination, and guessing affect the shape of the ICC.

Throughout this handbook, we have presented an assortment of graphs when introducing a measurement model and when presenting findings from a measurement model. In Chapter 2, we showed plots for describing continuous and categorical data; in Chapter 3, we used plots to show how reliability coefficients change as a function of different D studies; in Chapter 4, we used plots to determine how many factors we might be able to extract from the data and introduced a plot for identifying influential cases; and finally in Chapters 5 through 8, we used graphs to describe the functioning of items, tests, and information within the item response theory (IRT) framework. In this chapter, we expand on the potential for utilizing graphs within the context of these measurement models.

9.3   Diagnostic Plots

Statistical models are the core elements of educational measurement and psychometrics. All statistical models have assumptions, and the estimated relationships, represented by parameter estimates, are affected unequally by the data that go into the models. Therefore, assessing leverage, influence, and outliers is an essential exercise for all statistical modeling, and measurement models are no exception. One particularly helpful R package, the **faoutlier** package (Chalmers & Flora, 2015), implements several diagnostic and influential methods described in Flora, LaBrish, and Chalmers (2012) and the forward search algorithm described in Mavridis and Moustaki (2008). In the following section, we describe a few diagnostic tools that have been widely used for identifying influential observations and outliers in measurement models. As a first step, we install the **faoutlier** package and activate it along with the **hemp** package using the `library` command.

```
install.packages("faoutlier")
library("faoutlier")
library("hemp")
```

In Chapter 4, we introduced a plot that was described in Bollen (1989) for detecting multivariate cases with high leverage. An alternative approach is to plot the Mahalanobis distance.[1] Leverage is the measure of the uniqueness of a set of values on some variables (e.g., independent variables, dependent variables, or manifest variables). The more unique the combination of values for these variables in our data for a particular observation, the higher the leverage will be for that particular observation. The Mahalanobis distance can be defined as:

$$D_i^2 = (\mathbf{X}_i - \bar{\mathbf{X}})^T \mathbf{\Sigma}^{-1} (\mathbf{X}_i - \bar{\mathbf{X}}), \tag{9.1}$$

where $\mathbf{X}_i$ is a vector of variables for observation i, $\bar{\mathbf{X}}$ is a vector of means of these variables, and $\mathbf{\Sigma}$ is the sample covariance matrix of the variables. Conceptually, the Mahalanobis distance is the squared distance that an observation is from the multivariate centroid of the variable scaled by the covariance matrix.

The Mahalanobis distance can be calculated using the `mahalanobis` function or using the `robustMD` function in the **faoutlier** package. The `robustMD` function relies on the `cov.rob` function in the **MASS** package (Venables & Ripley, 2002) to calculate a robust estimate of the multivariate location and scale of the variables. In the context of factor analysis, manifest variables are typically assumed to be related to one another through their relationship with an underlying factor. If an observation has a high Mahalanobis distance, it implies that at least one of the values for a manifest variables is unusual relative to the values on the other manifest variables and this observation could result in a poor model fit or estimation errors.

Below we calculate the Mahalanobis distance for the cognitive measures from **interest** data set (we save the columns 4 through 9 in **interest** as `cognitive` for this example) with the `robustMD` function, save the results as D2, and finally plot them.

```
cognitive <- interest[, c(4:9)]
D2 <- robustMD(cognitive)
plot(D2, ylab = "D2")
```

In Figure 9.1, we see that one of the observations has a large value of the Mahalanobis distance. We can identify this observation using the following command:

---

[1]The Bollen's plot is based on a rescaled version of the Mahalanobis distance.

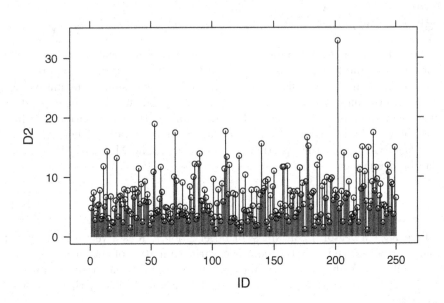

FIGURE 9.1
Mahalanobis distance plotted against the observation identifier.

```
D2$mah[which.max(D2$mah)]
```

```
    202
32.8935
```

The output shows that the observation with the highest Mahalonobis distance is 202 (i.e., person ID is 202). Figure 9.1 shows that the D^2 for this observation is nearly twice as large as the nearest observation's distance. We examine this observation further by printing the values associated with this observation as well as the maximum value for each variable using the `apply` function.

```
cognitive[202,]
```

	vocab	reading	sentcomp	mathmtcs	geometry	analyrea
202	2.63	2.23	2.55	1.38	3.86	3.5

```
apply(cognitive, 2, max)
```

vocab	reading	sentcomp	mathmtcs	geometry	analyrea
2.63	2.70	2.73	3.06	3.86	3.50

We see that this observation has the maximum value for vocab, geometry, and analyrea, and large (but not the maximum) values for reading, sentcomp, and mathmtcs. This is the likely reason for the high D^2. A major issue with assessing leverage through the Mahalanobis distance is that leverage is not necessarily a measure of influence and not based on a fitted model but instead based solely on the relationship of the manifest variables. A more useful statistic would take into account the model being fit and the effect of the parameters on the estimated fit and parameters. This is referred to as influence. The most common statistics for assessing influence are Cook's distance (Cook, 1977) and the generalized Cook's distance (Pek & MacCallum, 2011). These statistics can be calculated and used with measurement models estimated with **sem** (Fox et al., 2017), **lavaan** (Rosseel, 2012), and **mirt** (Chalmers, 2012) (i.e., many of the models discussed throughout this book). The generalized Cook's distance is defined as

$$gD_i = (\hat{\theta} - \hat{\theta}_{(i)})^T [\text{V\^{A}R}(\hat{\theta}_{(i)})]^{-1} (\hat{\theta} - \hat{\theta}_{(i)}), \tag{9.2}$$

where $\hat{\theta}$ and $\hat{\theta}_{(i)}$ are vectors of parameter estimates obtained from the full data set and from the data set with observation i removed, and $\text{V\^{A}R}(\hat{\theta}_{(i)})$ is a matrix containing the estimated asymptotic variances and covariances of the parameter estimates obtained with observation i removed. Observations with large generalized Cook's distance indicate strong influence on the parameter estimates in the full data set.

The generalized Cook's distance can be calculated with the gCD function in the **faoutlier** package. The gCD function takes a data set and an exploratory or confirmatory factor model (created with either the **sem**, **lavaan**, or **mirt** packages) and returns the generalized Cook's distance.

In the following example, we demonstrate the generalized Cook's distance by fitting a two-factor model to the **interest** data set and then computing the generalized Cook's distance for the observations in the **interest** data set. Below we describe the two-factor model for this example in more detail.

Vocab $= \lambda_{1,1} f_1 + \epsilon_1$	Mathmtcs $= \lambda_{4,2} f_2 + \epsilon_4$
Reading $= \lambda_{2,1} f_1 + \epsilon_2$	Geometry $= \lambda_{5,2} f_2 + \epsilon_5$
Sent Comp $= \lambda_{3,1} f_1 + \epsilon_3$	Analyrea $= \lambda_{6,2} f_2 + \epsilon_6$

In the models presented above, the λs are the standardized factor loadings, the fs are the factors, and the ϵs are the residuals (uniqueness). The first factor, f_1, might be called verbal intelligence and the second factor, f_2, mathematical reasoning. The expectations of the f_1 and f_2 are both 0 and the covariance matrix and the variance of the errors, $Var(\epsilon_j)$, are given by:

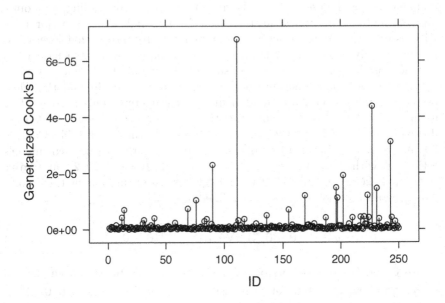

FIGURE 9.2
Generalized Cook's distance plotted against observation ID.

$$Var \begin{pmatrix} f_1 \\ f_2 \end{pmatrix} = \begin{pmatrix} 1 & \phi_{f_1,f_2} \\ \phi_{f_1,f_2} & 1 \end{pmatrix} \text{ and } Var(\epsilon_j) = \psi_j \mathbf{I}, \qquad (9.3)$$

where is the **I** is the identity matrix and the variances are fixed at 1 to identify the model and obtain standardized estimates. This model is defined in **lavaan** using the following commands:

```
cog_mod <- "
verb =~ vocab + reading + sentcomp
math =~ mathmtcs + geometry + analyrea
"
```

Below we calculate the generalized Cook's distance using the gCD function, save the results as **gd**, and finally plot them.

```
gd <- gCD(interest, cog_mod)
plot(gd, ylab = "Generalized Cook's D")
```

The results in Figure 9.2 show that one of the observations has a large value of the generalized Cook's distance. As we have done for the Mahalanobis distance, we can see the observation with the largest generalized Cook's distance by using the following command:

```
gd$gCD[which.max(gd$gCD)]
```

```
        111
6.765384e-05
```

The output shows that observation 111 has the largest generalized Cook's distance and is 1.5 times larger than the nearest observation (227). The generalized Cook's distance and Mahalanobis distance did not identify the same observations. This is unsurprising as the generalized Cook's distance is model dependent, while the Mahalanobis distance is not.

Table 9.1 shows the parameter estimates (i.e., factor loadings, factor covariance, and fit indices) and fit statistics for this model with observation 111 (in the third column) and without observation 111 (in the last column). When observation 111 is included, the model has good fit (i.e., high TLI and CFI and low RMSEA) and the manifest variables load highly on their respective factors. The model fit statistics and parameter estimates change little without observation 111 (identical to the hundredths place) and statistical significance is not affected. Therefore, while this observation has relative large influence, its removal from the model has no substantial impact.

Observations with high influence or leverage should not just be dropped from a data set. It is important to understand the cause of high influence or leverage. Perhaps it was caused by a data entry issue, a part of the population that has been undersampled, or caused by some other sampling/design issue. Either way, justifications are essential before removing observations and subjects, and influence and leverage should be used as additional tools to assess statistical validity.

TABLE 9.1
Standardized parameter estimates from the cognitive model using the full `interest` sample and without observation 111.

	Parameter	Full Sample	No 111
Vocab	$\lambda_{1,1}$	0.947	0.949
Reading	$\lambda_{2,1}$	0.850	0.845
Sentcomp	$\lambda_{3,1}$	0.856	0.851
Mathmtcs	$\lambda_{4,2}$	0.933	0.931
Geometry	$\lambda_{5,2}$	0.824	0.819
Analyrea	$\lambda_{6,2}$	0.876	0.872
$Cov(f_1, f_2)$	ϕ_{f_1,f_2}	0.807	0.800
RMSEA		0.051	0.052
CFI		0.996	0.996
TLI		0.992	0.992

If we want to ascertain the plausibility of the normality assumption underlying factor analysis, we can construct a normal Q–Q plot for each manifest variable. Just like in linear regression, this will let us know if a transformation may be necessary to achieve normality. The following code using
the obs.resid function in the **faoutlier** package computes the observational
residuals (Bollen & Arminger, 1991; Jöreskog, Sörbom, & Wallentin, 2006) and
prints a normal Q–Q plot for each of our manifest variables in the cog_mod object. The standardized forms of the observational residuals returned from the
obs.resid function are calculated using the procedures described in Bollen
and Arminger (1991). Below we extract the standardized observational residuals, reshape the data into a long format, and use the qqmath function in the
lattice package to create a normal Q–Q plot. The resulting normal Q–Q plot
is shown in Figure 9.3.

```
cog_resid <- obs.resid(interest, cog_mod)
cogn_std <- data.frame(cog_resid$std_res)
cogn_std_1 <- reshape(cogn_std,
                      direction = "long",
                      varying = 1:6,
                      v.names = "std_res",
                      timevar = "var",
                      times = names(cogn_std))
qqmath(~ std_res | var, cogn_std_1,
       xlab = "Normal theoretical quantiles",
       ylab = "Standardized Residuals",
       panel = function(x, ...){
         panel.qqmathline(x, ...)
         panel.qqmath(x, ...)
         }, auto.key = T)
```

In Figure 9.3, the verbal manifest variables are on the top row and the
mathematics manifest variables are on the bottom row. This plot shows that
the normality assumption is reasonable for each manifest variable because the
standardized residuals are mostly lined up on the diagonal line in the normal
Q–Q plots.

There are a couple of issues with using the normal Q–Q plot with the
standardized observational residuals. First, the normal Q–Q plot will not necessarily pick up a non-linear relationship between the factor and the manifest
variable. Second, the standardized observational residuals for a manifest variable will be normally distributed if that manifest variable is unrelated to the
factor and the manifest variable is marginally normally distributed.

To better ascertain the plausibility that the relationship between the manifest variable and the factors are in fact linear, we can plot each manifest
variable against the estimated factor scores (Jöreskog et al., 2006) or the
standardized observational residuals against the estimated factor scores. We

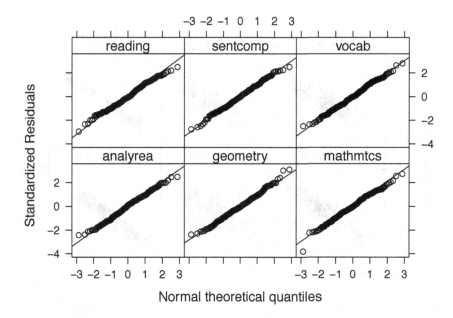

FIGURE 9.3
Normal Q–Q plot comparing standardized residuals for the six manifest variables against normal theoretical quantiles.

generally favor the former method because traditional residual plots show no relationship between the residuals and the predicted values when the model is correctly specified. If one plots the observational residuals against the estimated factor scores (a worthy exercise), then the resulting relationship visible in the plot is indicative of the relationship between the factor and the manifest variables and will be flat and random, not if the model is correctly specified, but if there is no relationship.

In order to do create this plot, we first extract the estimated factor scores from a `lavaan` object using the `predict` function and convert them to a data.frame called `cog_pred`.

```
cog_fit <- cfa(cog_mod, interest)
cog_pred <- data.frame(predict(cog_fit))
```

Next, we reshape the `cognitive` data set into a long format, combine it with the estimated factor scores, and finally create our plot with the `xyplot` function. Figure 9.4 shows the output of this procedure. In the top row, the verbal manifest variables are plotted against the estimated verbal factor scores, while in the bottom row the mathematics manifest variables are

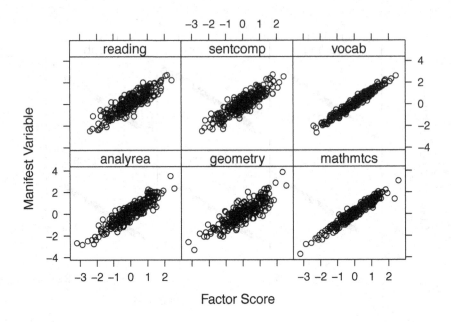

FIGURE 9.4
Manifest variables plotted against the estimated factor scores.

plotted against the estimated mathematics factor scores. The relationship between the manifest variables and the factor scores in Figure 9.4 is strongly positive and linear (corroborated by the large factor loadings). If the relationship between the factor and the manifest variables were not linear, then this would show up as a non-linear pattern in the plot.

```
cognitive_1 <- reshape(cognitive,
                       direction = "long",
                       varying = 1:6,
                       v.names = "obs_score",
                       timevar = "var",
                       times = names(cognitive))
cognitive_1$pred_fact <- c(rep(cog_pred$v, 3),
                           rep(cog_pred$m, 3))
xyplot(obs_score ~ pred_fact | var, cognitive_1,
       xlab = "Factor Score",
       ylab = "Manifest Variable",
       auto.key = T)
```

Because we have access to the standardized residuals, we can also ascertain if there are certain observations that are large and do not fit our proposed

Observed Residuals

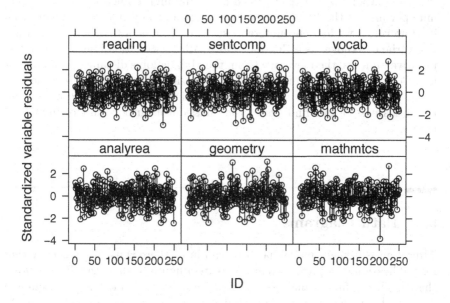

FIGURE 9.5
Standardized residuals against ids by the six manifest variables.

model. We demonstrate this below by plotting the standardized observational residuals against their identifiers (the result is shown in Figure 9.5).

```
plot(cog_resid, restype = "std_res")
```

We see that, in general, the residuals are all within a reasonable range (i.e., between -3 and 3). We can quickly identify which observations fall outside of this range using the `which` function. The code below identifies and prints the observations where the absolute value of the standardized residuals is greater than 3.

```
outliers <- which(abs(cogn_std_1$std_res) > 3)
cogn_std_1[outliers, ]
                 var    std_res  id
202.mathmtcs mathmtcs -3.810076 202
114.geometry geometry  3.032828 114
202.geometry geometry  3.120521 202
```

We see that observation 202 has a standardized residual greater than 3 for mathmatcs and geometry and observation 114 has a standardized residual greater than 3 for geometry. Observation 212 was the same observation identified by the Mahalanobis distance, while 114 was not. Given that these two observations were not identified as having high influence, they are likely unimportant to the fit of the overall model and any inferences made from it. However, if we had encountered issues during model fitting, such as negative variance (known as a Heywood case), other estimation issues, or counterintuitive findings, then leverage (from the Mahalanobis distance), the generalized Cook's distance, and investigating the residuals would be a great place to begin our detective work. In addition, it is always a good idea to investigate the robustness of our findings to the omission of outliers and/or influential points.

9.4   Path Diagrams

After a measurement model has been run in R, there is usually a need, either for publication or a presentation, to create a figure to show the findings from the model. For factor analysis and structural equation models, it is common to show the model findings via path diagrams. There are a myriad of ways to create path diagrams (including Microsoft PowerPoint, Ωnyx (von Oertzen, Brandmaier, & Tsang, 2015), or Adobe Photoshop). Although there are several R packages capable of producing high-quality path diagrams, the **semPlot** package (Epskamp & Stuber, 2017) is arguably the most user-friendly package. The `semPaths` function within the **semPlot** package can take a **lavaan** object directly and automatically create a path diagram.

Below we demonstrate how to create a path diagram using the `semPaths` function. We begin by installing the **semPlot** package and loading it.

```
install.packages("semPlot")
library("semPlot")
```

The R code below creates the path diagram shown in Figure 9.6. The code tells `semPaths` to plot the estimated standardized parameters; to suppress plotting residuals of the manifest variables and the exogenous latent variables' variances; to rotate the plot 90 degrees from the default orientation (which is top-down); and to change the color schemes to black (`color = "black"`, `edge.color = "black"`) and size of the fonts for the number on the paths (`edge.label.cex = 1`).

```
semPaths(cog_fit, what = "std", residuals = F,
         rotation = 4, color = "black",
         edge.color = "black",
         edge.label.cex = 1)
```

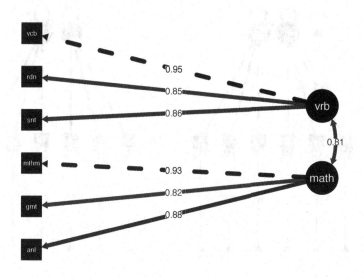

FIGURE 9.6
Path diagram for the two-factor cognitive model.

In Chapter 11, we present measurement invariance and how to run multi-group models with **lavaan**. The semPaths function can easily create a multi-group path diagram for a factor model estimated with **lavaan**. Below we fit our multi-group configural model for gender for our cognitive two-factor model (this configural model is discussed in detail in Chapter 11) and then use the semPaths function to quickly create a path diagram. From Figure 9.7, we can quickly and easily compare the factor loadings, intercepts, and errors between these two groups (left: female, right: male).

```
configural <- cfa(cog_mod, interest, group = "gender")
semPaths(configural, what = "est",
        edge.label.cex = 1, panelGroups = T,
        ask = FALSE, fade = FALSE, title = FALSE,
        color = "black", edge.color = "black")
```

The semPaths function can also be used to easily plot other models, including structural equation models, path analysis, and regressions, while still affording the high level of customization necessary for making publication-worthy figures.

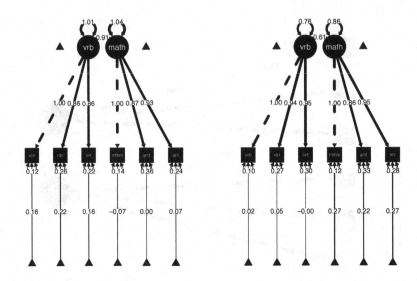

FIGURE 9.7
Path diagram of the multi-group, configural model for the two-factor cognitive model.

9.5   Interactive Plots with shiny

The **shiny** package (Chang et al., 2017) is an interactive web-application framework for R. It can be used to create some very impressive, interactive graphics and dashboards in R with little or no HTML, CSS, or JavaScript required.[2]

The exact details of **shiny** are beyond the scope of this chapter and this book. Instead of a detailed explanation of **shiny**, we focus on how to quickly create two basic **shiny** applications and to demonstrate how **shiny** can be used to further understand measurement models. As there have been several books written on **shiny**, as well as excellent tutorials on the **shiny** website,[3] we refer our readers to these resources if our applications have piqued their interest and they have more advanced needs.

[2]See https://shiny.rstudio.com/gallery/ for examples.
[3]https://shiny.rstudio.com/tutorial

An important thing to note is that there are a couple of different ways to create a **shiny** application. In the following examples, we demonstrate how to create a **shiny** application within a single R script. There are pros and cons of using a single script, but, in general, we recommend using a single script for small applications. To create a **shiny** application, we need to define two functions:

1. A `ui` function: This function will create the user interface (e.g., the sliders, check boxes, and other design features of the application).

2. A `server` function: This function will take the input from the `ui` function and create plots, text, summaries, etc., based on the input.

These two functions are then passed to the `shinyApp` function in the **shiny** package.

9.5.1 Example 1: Diagnostic Plot for Factor Analysis

For our first **shiny** application, we will create an interactive diagnostic plot of the manifest variables against the estimated factor scores. This is the plot that we described earlier in this chapter. This application will allow us to understand how violating key assumptions of factor analysis (e.g., linear relationship between the manifest variable and the factor and the condition that a relationship between the manifest variable and the factor exists) are manifested in this plot. By creating this **shiny** application, we will have a better understanding of what this plot can tell us.

In our application, the user will be able to specify no relationship (by setting the value of `lam` to 0); that the relationship between the manifest variable and the factor is a linear relationship; or that the relationship is a nonlinear relationship (by checking the checkbox). The script for this application is below.

Within the script, we comment on what each part of the code is doing.

```
# Load the required R packages
library("shiny")
library("lavaan")


# Shiny ui's can be drawn using a grid system
# consisting of 12 columns wide. We make use of this below
# by using the first 4 columns to contain the inputs (widgets)
# and the remaining 8 columns to hold the plot.

ui <- fluidPage(
  column(4,
```

```
# This creates the input for setting our factor loading
# for manifest variable one. The default value is set to
# 0.5 and if we click the arrows, it will increase/decrease
# by 0.5
numericInput("lam",
  "Value for the factor loading for manifest variable 1.",
  value = .5,
  step = .5),

# This creates the checkbox for a non-linear
# quadratic relationship. When the box is unchecked,
# the relationship is linear.
checkboxInput("nonlinear",
  "Should manifest variable 1 have a non-linear
    relationship with the factor?",
  value = F)
),
column(8,
# This will draw our diagnostic plot
plotOutput("diagPlot")
)
)

# Below we define a server function
# Every time we refresh the input random
# data will be created. So, no two plots will
# be identical!

server <- function(input, output) {
  output$diagPlot <- renderPlot({

  # Create our factor as a standard random normal variable.
  # rnorm() ensures every time the plots are different.
  f1 <- rnorm(500)

  # If we specify the non-linear relationship
  # then f1 will be squared else it won't be.
  # We also add some specific variance/uniqueness
  # (the rnorm part at the end).
  if(input$nonlinear) {
    m1 <- input$lam * f1 + 2 * f1^2 + rnorm(500)
  } else {
    m1 <- input$lam * f1 + rnorm(500)
  }
```

```
# Now do the same thing for manifest variables 2 - 4,
# except their relationship with the factor is always
# fixed.
# The .60, .75, and .9 are the factor loadings
# The rnorm() adds specific/unique variance
m2 <- .60 * f1 + rnorm(500)
m3 <- .75 * f1 + rnorm(500)
m4 <- .9 * f1 + rnorm(500)
dat <- data.frame(m1, m2, m3, m4)

# Define the CFA model, run it, and extract the factor scores
mod <- '
  fact =~ m1 + m2 + m3 + m4
  '
fit <- cfa(mod, dat, std.lv = T)
dat$pred <- predict(fit)[,1]

# Create a 2 by 2 plot (the par stuff) and
# add a blue LOWESS curve to each of the manifest
# variables against the est. factor scores sub plots.
par(mfrow = c(2,2))
for(i in 1:4){
  plot(dat[,i] ~ pred, dat, xlab = "Est. Factor Score",
       ylab = paste("Manifest Var", i))
  lines(lowess(dat$pred, dat[,i]), col = "blue")
}
})
}
```

Once the `ui` and **server** functions are defined in R, then we can call the `shinyApp` function passing these functions as follows:

```
shinyApp(ui = ui, server = server)
```

Alternatively, this **shiny** application can be called from the **hemp** package by typing:

```
fadiag_demo()
```

A screenshot of this application is shown in Figure 9.8. As can be seen in Figure 9.8, the **shiny** application has two interactive inputs: a numeric input, which corresponds to the size of the factor loading for the first manifest variable, and a check box, which, when selected, specifies a non-linear relationship between the manifest variable and the factor. The manifest variables 2 through 4 are always linearly related to the factor to various degrees (see the above script for the size of these loadings). Only the relationship between

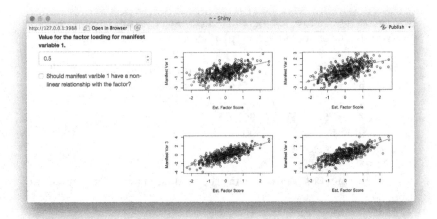

FIGURE 9.8
Screenshot of the **shiny** application for the factor analysis diagnostic plot.

manifest variable 1 and the factor, the upper-left plot, will change. However, each plot will look slightly different because we are randomly generating new data every time.

By altering the values in the numeric input, one can understand how the direction and magnitude of the factor loading, when the relationship is linear, manifests itself in this diagnostic plot; what no relationship looks like (*hint*: it will look like a well-behaved multiple regression residual plot), and by selecting the check box, one can understand what a non-linear relationship would look like. This application, with some slight modifications, could be made to work with the normal Q–Q plot that we have described earlier in this chapter. This is left as an exercise for keen and curious readers.

9.5.2   Example 2: The 3PL IRT Model

Our second **shiny** application will allow us to explore how the shape of the item response function of the 3PL IRT model changes when item difficulty, item discrimination, and guessing are manipulated. As with the first **shiny** application, our comments about what the code does are embedded within the R script below.

```
# Load the shiny package
library("shiny")

# Define the 3PL function
```

```
threepl <- function(person, b, a, c) {
  x <- c + (1 - c)*(exp(a * (person - b)) /
                    (1 + exp(a * (person - b)))))
  return(x)
}


# Define the range of abilities, difficulties,
# discriminations, and guessing parameters allowed
ability <- seq(-3, 3, by = .1)
diff <- -3:3
discr <- -2:2
guess <- 0:10/10


# Create a data.frame with all the combinations
# of the values permitted above
parameter_setup <- expand.grid(person = ability,
                               b = diff, a = discr, c = guess)
# Run the threepl() function and save the probabilities
# of getting an item correct as function of these parameters
parameter_setup$prob <- threepl(person = parameter_setup$person,
                                b = parameter_setup$b,
                                a = parameter_setup$a,
                                c = parameter_setup$c)


# Define the user-interface (ui) function
# Will consists of three inputs
#
# - A slider to set item difficulty (b)
# - A slider to set item discrimination (c)
# - A slider to set guessing (a)
#
# and the plot (called threepl).
#
# Note: when a = 1 and c = 0, this the Rasch
# and when c = 0, this is the 2PL
#
# Shiny ui's can be drawn using a grid system
# consisting of 12 columns wide.We make use of this below
# by using the first 4 columns to contain the widgets
# and the remaining 8 columns to hold the plot.
#
ui <- fluidPage(
  column(4,
  sliderInput("b", label = "Item Difficulty",
```

```
                    min = -3, max = 3, value = 0, step = 1),
    sliderInput("a", label = "Item Discrimination",
                    min = -2, max = 2, value = 1, step = 1),
    sliderInput("c", label = "Guessing",
                    min = 0, max = 1, value = 0, step = .1)),
    column(8,
    plotOutput("threepl"))
)


# Below we define a server function that
# subset our data based on the values of the sliders
# and then plot the IRF.
server <- function(input, output){
    output$threepl <- renderPlot({
        plot.data <- subset(parameter_setup, b == input$b &
                            a == input$a &
                            c == input$c)
        plot(prob ~ person, plot.data, type = "l",
            xlab = expression(theta),
            ylab = "Pr(Y = 1)",
            ylim = c(0, 1))
    })
}
```

Once again, we call the ui and **server** functions using the **shinyApp** function as follows:

```
shinyApp(ui = ui, server = server)
```

Alternatively, this **shiny** application can be called from the **hemp** package by typing:

```
threepl_demo()
```

The default view of this application is presented in Figure 9.9. In our **shiny** application, we have three slider inputs that we can interact with: item difficulty, item discrimination, and guessing. The astute reader will note that the default view is actually the one-parameter logistic (1PL) IRT model, and more specifically, the Rasch model.

With this application, we can also learn about the two-parameter logistic (2PL) model in addition to the Rasch, 1PL, and 3PL models. When we fix the guessing parameter to 0 and item discrimination to a specific value and only alter the item difficulty, then we have the 1PL model (when item discrimination is fixed to 1, we have the Rasch model). When we fix the guessing parameter to 0 and alter the item discrimination and difficulty, then we have the 2PL model. Finally, if we alter all three parameters, then we have the 3PL

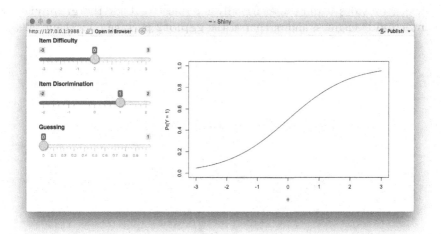

FIGURE 9.9
Screenshot of the **shiny** application for the IRT plot.

model. This application will allow us to understand what negative discrimination or no discrimination would look like, how increasing guessing can alter the possible range of probabilities, and so on.

9.6 Summary

In this chapter, we presented three very different uses of graphics and visualization in measurement. Namely, we used graphics for diagnostics and understanding the appropriateness of our model for our data; for creating path diagrams for presentations and publications; and for learning about our diagnostic plots and our models. While we could not cover these topics in too much detail, we hope that our readers will have an increased appreciation for the use of visualization in measurement; some ideas for how they might use graphics with their data; and some code to help them get started creating their own **shiny** applications and publication-ready figures. For those readers with specific, advanced visualization needs or who have a specific interest in data visualization, we recommend that they check out the CRAN Task View for Graphic Displays;[4] the **DiagrammeR** package (Sveidqvist et al., 2017),

---

[4]https://cran.r-project.org/web/views/Graphics.html

which provides a more user-friendly interface to R for the **graphviz**[5] and **mermaid**[6] engines; and of course, the **ggplot2** package (Wickham, 2009).

---

[5]http://www.graphviz.org/
[6]https://knsv.github.io/mermaid

10

Equating

10.1 Overview

This chapter introduces the test equating methods without using item response theory (IRT). The chapter begins with a brief introduction to equating. This includes a review of the various equating and linking designs (single group, equivalent group, and non-equivalent group) and functions (general linear and equipercentile). Examples using different designs and functions are demonstrated using the **equate** package (Albano, 2016) in R.

10.2 Introduction

Students in their third year of high school in the United States typically take a college admission test. These admission tests are administered on multiple test dates within a calendar year to provide the examinees with some flexibility in selecting their test date and to allow them to retake the exam. Because the tests are administered multiple times within and across calendar years, different test questions are needed. If the same test questions were used across multiple occasions, then examinees who took the test earlier could inform later examinees about the content and answers to the items, or an examinee who took the exam multiple times to achieve a desired score might get a higher score, not because of an increased knowledge of the construct(s), but instead because of prior exposure to the questions or an increased ability to take the test. Therefore, instead of using the same exam, different forms of the exam are constructed based on the same test specifications, often with the help of a test blueprint. This helps to ensure that the test forms are similar and scores across different forms are comparable.

Because of concerns about the security of the test, namely issues surrounding item exposure, a statistical procedure known as equating is commonly used to make adjustments on test scores from different test forms. Why do the scores need to be adjusted? Because one of the test forms may be more difficult than another, and examinees taking a particular test form are expected to have lower scores because of the difficulty of the instrument and not because their knowledge of the tested construct(s) is less. This is a type of measurement error and it is correctable through the equating procedure. Equating adjusts for the variation in difficulty of the test forms and allows the scores on the various test forms to be interchangeable and more comparable.

According to Kolen and Brennan (2014), there are several desirable properties of equating:

1. The equating transformation is symmetric (symmetry property).

2. The alternate forms are built to have the same content and statistical specifications (same specifications property).

3. The test form that examinees receive does not matter (Lord's equity property).

4. The equating relationship is the same regardless of the group of examinees used for equating (group invariance property).

5. The characteristics of score distributions are set equal for a particular population of examinees (observed score equating properties)

Kolen and Brennan (2014) also present a useful list of steps for implementing test equating. This list includes identifying a purpose for equating; constructing alternate forms of the exam; considering the equating designs and operational definition; estimation; and finally, evaluating the findings. In this chapter, we focus on the steps involving the equating design, estimation methods, and evaluation of the results from equating, as this has ramifications for how the equating is performed in R.

10.2.1   Equating Designs

Different equating designs can be employed in an equating study and the decision of which design to use has important statistical implications. Examinees involved in the equating study should be representative of the population of examinees, and the study should be conducted under typical test settings. To follow the convention used in the **equate** package that we are going to use later in the chapter, we categorize equating designs as either a single group, equivalent groups, or nonequivalent groups.

In a single group design, a random sample of examinees takes both forms of an exam, X and Y. Because of test fatigue and potential effects of ordering (e.g., increased performance due to increased familiarity with items), counterbalancing may be considered. Counterbalancing involves a portion of the examinees taking form X first (subgroup A) and the other portion taking form Y first (subgroup B). Any differences in the distribution of scores on forms X and Y are assumed to be caused by differences in test form difficulty, whereas if counterbalancing was not done, then differences between X and Y could be caused by other factors, such as fatigue. For example, if the mean on test form X is 10 points and the mean on test form Y is 13 points, then test form Y is 3 points easier than form X (assuming that the ordering had no effects).

In an equivalent group design, a random sample of examinees takes form X and a second random sample takes form Y. Because these are random samples from the same population, any differences in the distribution of scores are again assumed to be a function of test difficulty because the randomization should ensure that the two groups are equivalent in the ability being assessed. If the group taking form X had a mean of 10 and the group taking Y had a mean of 13, then test form Y would again be 3 points easier than form X.

In a nonequivalent group design, the population of examinees taking the different forms cannot be assumed to be from the same population. Therefore, potential differences in their abilities on the construct(s) must be accounted for. Differences in abilities are accounted for through the use of common items (i.e., anchor test). Scores on these common items may either be internal (contributing to the examinee's score on a particular test form) or external (not contributing to the examinee's score on a particular test form).

10.2.2 Equating Functions and Methods

An equating function is used with a single group or equivalent group design to place the test forms onto a common scale. If, however, a nonequivalent group design is used, then the techniques used to place the forms on a common scale are referred to as equating methods. The distinction seems subtle but again has important statistical implications. With an equating function, the examinees are assumed to come from the same population, while an equating method assumes that the examinees come from two different populations (e.g., P1 and P2). Because of potential differences between P1 and P2, an equating method is necessary to construct a *synthetic* population, which represents a weighted mixture of these two populations using the common and unique items. The selected equating method relies on assumption about the relationship between the total scores and the common-item scores for P1 and P2 (Braun & Holland, 1982; Kolen & Brennan, 2014).

The equating functions in the **equate** package fall into two broad categories: linear and nonlinear functions. The linear functions include identity, mean, linear, and general linear functions. The nonlinear functions include equipercentile, circle-arc, and composite functions. There are also several equating methods available in **equate**. They are the nominal weights,

Levine, Tucker, Braun, frequency estimation, and chained equating methods. These methods can be used in conjunction with several of the functions listed above (see Table 1 in the **equate** vignette and Table 8.5 in Kolen and Brennan (2014) for additional help in selecting an appropriate equating function). These functions and methods will be described briefly when they are presented below.

10.2.3   Evaluating the Results

After equating has been performed, the results should be evaluated. This may involve calculating standard errors of equating through bootstrapping, formally assessing the aforementioned properties, or assessing the consistency of the results. Here the focus is on fit measures that can be obtained from the **equate** package. Standard errors, bias, and root mean square error obtained via bootstrapping, as well as visually comparing the results from different equating functions/methods, can be used as a sensitivity analysis. Chapter 8 in Kolen and Brennan (2014) presents the procedures for evaluating equating in more detail.

10.2.4   Further Reading

Given the brevity of this chapter, for readers requiring a gentler introduction to equating, we encourage checking out the "Equating Test Scores (Without IRT)" paper by Livingston (2014). Readers interested in a deeper, more through treatment of equating are strongly referred to Kolen and Brennan (2014). Finally, readers with alternative equating designs who want to use other equating functions within the **equate** package should read the article prepared by Dr. Anthony Albano (the author of **equate**) appearing in the *Journal of Statistical Software* (Albano, 2016). An unpublished version of this article can also be retrieved by typing the following in the R console (after installing and activating the **equate** package).

```
vignette("equatevignette")
```

10.3   Examples

Before we begin the equating examples, we first install the **equate** package and activate it along with the **hemp** package using the `library` command.

```
install.packages("equate")
library("equate")
library("hemp")
```

10.3.1 Equivalent Groups

10.3.1.1 Identity, Mean, and Linear Functions

Identity and mean functions are all restricted variants of the linear function. The linear function, in turn, is a special case of the general linear function (Albano, 2015). These functions all differ based on the adjustments that they make in order to place one form onto the same scale of another form. If we want to relate the scores from form X (the new form) to the scores on form Y (the reference form), then we can use the following functions:

- The *identity* function would make no adjustment and assumes that X and Y have the same scale properties (i.e., distribution properties). X is already on the same scale as Y by virtue of it having identical properties to Y.

- The *mean* function adjusts the scores by the difference in means between X and Y. If X is harder than Y, then this difference is added to the scores on X; if X is easier than Y, then this difference is subtracted from the scores on X.

- The *linear* function adjusts the scores based on the means and standard deviations of test forms X and Y. The scores are equated such that if a test score was one standard deviation above the mean on the X form, the equated score on Y would be one standard deviation above the mean.

To demonstrate these functions, we use the `hcre` data set from the **hemp** package. The `hcre` data set contains the score distributions for two forms of a hypothetical college readiness exam. Potential raw scores on these forms range from 20 to 50, and the two test forms, x and y, were each administered to 1000 students.

The **equate** package analyzes score distributions as frequency tables. These frequency tables are created using the `as.freqtab` function in **equate**. The data in `hcre` must be restructured as a frequency table in order to work with the functions in **equate**, which we do below as:

```
hcre_data <- as.data.frame(table(hcre$score, hcre$form))
names(hcre_data) <- c("total", "form", "count")
hcre_x <- as.freqtab(hcre_data[hcre_data$form == "x",
                               c("total", "count")],
                     scales = 20:50)
hcre_y <- as.freqtab(hcre_data[hcre_data$form == "y",
```

```
                       c("total", "count")],
            scales = 20:50)
```

The above R code creates a two-way contingency table using the `table` function, re-formats it as a data frame using the `as.data.frame` function, and then creates two separate frequency tables based on the test form each examinee was administered using the `as.freqtab` function. These frequency tables can then be summarized using the `summary` function and a bar plot created using the `plot` function.

```
rbind(form_x = summary(hcre_x), form_y = summary(hcre_y))
```

```
          mean       sd       skew      kurt min max    n
form_x 34.925 4.098069 0.04373977 3.169021  23  48 1000
form_y 38.840 3.815883 0.05785096 2.870284  28  50 1000
```

The summary function returns the mean, standard deviation, skewness, kurtosis, minimum, maximum, and number of observations for each form. The two forms have very similar sample statistics, with the exception that the mean is higher for form Y than form X (i.e., form X was the more difficult exam). This suggests that only a mean adjustment might be necessary.

```
plot(hcre_x)
```

Based on Figure 10.1, the distribution of the scores on form X seems to be unimodal, symmetric, and roughly normal. The distribution of form Y is very similar to that of form X except that it is slightly shifted to the right (not shown here). To place forms X and Y on the same scale, the `equate` function is used. To specify the function used in equating, the `type` argument is used. The following call uses the mean function in the equating process, saves the results to the object `mean_yx`, and prints the output.

```
mean_yx <- equate(hcre_x, hcre_y, type = "mean")
mean_yx
```

```
Mean Equating: hcre_x to hcre_y

Design: equivalent groups

Summary Statistics:
     mean   sd skew kurt   min   max    n
x   34.92 4.10 0.04 3.17 23.00 48.00 1000
y   38.84 3.82 0.06 2.87 28.00 50.00 1000
yx  38.84 4.10 0.04 3.17 26.91 51.91 1000

Coefficients:
intercept      slope        cx         cy         sx         sy
```

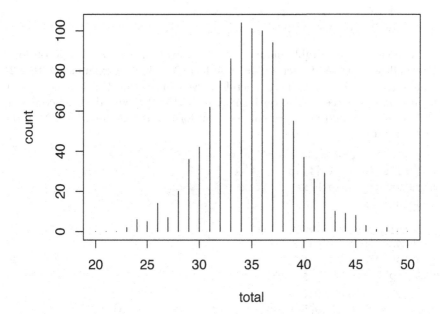

FIGURE 10.1
Bar plot of the test scores on form X in the hcre data set.

3.915	1.000	35.000	35.000	30.000	30.000

By default, the `equate` function returns the summary statistics for the new form (x), the reference form (y), and the form x after it has been put on the y scale (yx) followed by the estimated coefficients used in the equating process. If we compare the summary statistics for form X to those on the rescaled form X (yx), we see that the moments of the distributions are the same except for the mean, where the rescaled form X has the mean of form Y. This is because we used the mean function, which adds a constant to all the scores from form X. The difference between the X and Y forms is 3.915, which is the intercept coefficient reported in the output. So, each score on form X is 3.915 points higher. This can be seen by printing the concordance table as this table relates the scores on the new form to the reference form.

```
head(mean_yx$concordance)
```

```
  scale     yx
1    20 23.915
2    21 24.915
3    22 25.915
4    23 26.915
```

```
5     24 27.915
6     25 28.915
```

The concordance table indicates that an examinee who got a 20 on form X would be expected to get a 23.915 on form Y, while an examinee with a 21 would get a 24.915, and so on. Therefore, we can see that this transformation is just an examinee's score on form X + 3.915. We can add the score that the examinee would be expected to get on form Y if they took form X to the hcre_x data set.

```
form_yx <- mean_yx$concordance
colnames(form_yx)[1] <- "total"
hcre_xy <- merge(hcre_x, form_yx)
head(hcre_xy)
```

```
  total count     yx
1    20     0 23.915
2    21     0 24.915
3    22     0 25.915
4    23     2 26.915
5    24     6 27.915
6    25     5 28.915
```

To place the scores from form X onto form Y using linear equating, we need to substitute function = "mean" with function = "linear".

```
linear_yx <- equate(hcre_x, hcre_y, type = "linear")
linear_yx
```

```
Linear Equating: hcre_x to hcre_y

Design: equivalent groups

Summary Statistics:
     mean   sd skew kurt   min   max    n
x   34.92 4.10 0.04 3.17 23.00 48.00 1000
y   38.84 3.82 0.06 2.87 28.00 50.00 1000
yx  38.84 3.82 0.04 3.17 27.74 51.01 1000

Coefficients:
intercept     slope        cx        cy        sx        sy
   6.3199    0.9311   35.0000   35.0000   30.0000   30.0000
```

There are two important differences between this output and the output from the mean function. First, the mean and standard deviation for the equated scores for form X are the same as those for form Y (unlike before, where only the mean was the same). Second, the intercept and slope are estimated. To understand how these coefficients are calculated, we need to standardize the scores on the forms X and Y, and then solve for Y (which will be our adjusted X score on form Y):

$$\frac{X - \bar{X}}{s_X} = \frac{Y - \bar{Y}}{s_Y}$$

$$Y = \frac{s_Y}{s_X}(X - \bar{X}) + \bar{Y}$$

$$Y = \underbrace{\frac{s_Y}{s_X}}_{\text{slope}} X + \underbrace{\bar{Y} - \frac{s_Y}{s_X}\bar{X}}_{\text{intercept}},$$

(10.1)

where $\bar{X}$ and $\bar{Y}$ are the means of form X and Y and s_x and s_y are the sample standard deviations of form X and Y. If we plug the values printed above into Equation 10.1, we arrive at the same intercept and slope coefficients.

10.3.1.2 Nonlinear Functions

Two nonlinear equating functions are available for **equate**: equipercentile and circle-arc functions. Equipercentile equating works by breaking up the scores on the X and Y forms into percentiles. Then percentiles of form X are matched to the percentiles on form Y. The scores that correspond to these matched percentiles are then the new adjusted scores. For example, if a score corresponds to the 15th percentile on form X, then the adjusted X score would be the score that corresponds to the 15th percentile on form Y. The circle-arc function is a restricted curvilinear method that is appropriate when working with small samples (Kolen & Brennan, 2014).

We will show the use of nonlinear functions with the `hcre` data. Because the data are contrived, there are no extreme outliers or other notable irregularities in the score distributions. However, if there were irregularities in the distributions of either test form, then *presmoothing* may be considered. Presmoothing is important when equipercentile equating is to be performed. The presmoothing acts to reduce the effects of sampling error on percentile and percentile ranks. Presmoothing is performed using the `presmoothing` function in **equate**. Several presmoothing functions are implemented in the **equate** package, including replacing scores below a threshold with their adjacent scores, and the use of polynomial log-linear smoothing (see Chapter 3 in Kolen and Brennan (2014) for a thorough and technical discussion of smoothing). Equipercentile equating can be performed using the following code:

```
equi_yx <- equate(hcre_x, hcre_y, type = "equipercentile")
```

As an informal sensitivity analysis or as a means to better understand the

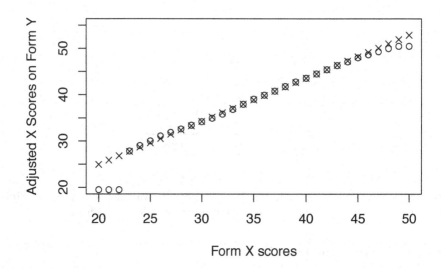

FIGURE 10.2
Scatterplot of the adjusted X scores on form Y against the original form X scores. (The hollow circles correspond to the equipercentile equating and the Xs correspond to linear equating.)

different equating functions, we can plot the adjusted scores from the different functions against the raw X scores. The code below does this by extracting the adjusted scores and the original scores from the concordances tables in the `mean_yx` and the `equi_yx` objects.

```
plot(equi_yx$concordance$yx ~ equi_yx$concordance$scale,
     type = "p", xlab = "Form X scores",
     ylab = "Adjusted X Scores on Form Y", ylim = c(20, 55))
points(linear_yx$concordance$yx ~ linear_yx$concordance$scale,
     pch = 4)
```

Figure 10.2 shows that the biggest discrepancies between these two functions are at the tails of the distribution. This is often the case as there is usually less mass at the tails of distributions. Figure 10.2 also shows the non-linear nature of the equipercentile equating (the hollow circles).

TABLE 10.1
Available equating functions for each equating method (adapted from Albano (2016)).

	Available Functions				
	Mean	Linear	General linear	Equipercentile	Circle-arc
Tucker	■	■	■		■
Nominal	■		■		■
Levine true score	■	■	■		■
Braun/Holland	■	■	■		■
Frequency				■	■
Chained	■	■	■		■

10.3.2 Nonequivalent Groups

When a single target population cannot be assumed, a nonequivalent group design is used and an equating method must be specified. The mathematics of these methods are beyond the scope of this chapter and interested readers are referred to Albano (2016) and Kolen and Brennan (2014) for more details. Generally, these methods work by relating the total scores on form X and form Y through the scores on a common set of items appearing on both forms (i.e., the common anchor scores) and the creation of a weighted, synthetic population. These methods differ in the manner in which they describe and estimate the form of this relationship. The Tucker, nominal weights, and Levine true-score methods rely on various forms of regression to do this, while frequency estimation and Braun/Holland do not. The chained equating method is the only method that does not explicitly involve the creation of a synthetic population.

An equating method must be specified in conjunction with an equating function. Table 10.1, adapted from Albano (2016), shows the equating functions available for each equating method. For each method, there are typically several functions that can be specified with the exception of the frequency method, which can only use the equipercentile function. In Table 10.1, the filled-in gray boxes indicate that the function is available for that method, and BH stands for the Braun/Holland method.

10.3.2.1 Linear Tucker Equating

The **negd** data set in **hemp** is used to demonstrate how to equate two forms that have been administered to separate populations. The data set **negd** contains data on 2,000 test takers responding to 35 questions on a particular form of the test (either form X or form Y, labeled **form** in **negd**). Questions 1 through 25, labeled q.1 ... q.25, are unique to each form, while questions 26 through 35, labeled a.1 ... a.10, are the common anchor items. Because

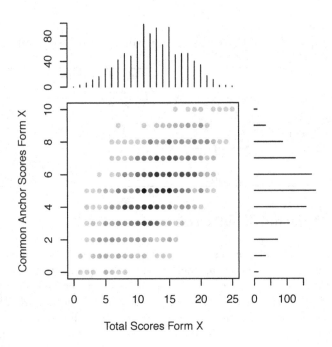

FIGURE 10.3
Plot of the common anchor scores against total scores on form X.

the data are in a person-by-item format, for each person, we need to calculate their total score on the form, excluding the common anchor items, then their score on the common anchor items. Once this is done, frequency tables can be created, plotted, and then passed onto the `equate` function.

```
negd$total <- rowSums(negd[, 1:25])
negd$anchor <- rowSums(negd[, 26:35])
negd_x <- freqtab(negd[1:1000, c("total", "anchor")],
            scales = list(0:25, 0:10))
negd_y <- freqtab(negd[1001:2000, c("total", "anchor")],
            scales = list(0:25, 0:10))
```

Figure 10.3 shows a scatterplot of the common anchor scores on form X against the total scores on form X. Plotted along the axes are the marginal distributions of these scores. The two distributions are roughly normal, with some irregularities for the total scores for form X (note the presence of various peaks near the center of this distribution). Because there are irregularities, assumed to arise from measurement error, presmoothing could be performed.

```
plot(negd_x, xlab = "Total Scores Form X",
     ylab = "Common Anchor Scores Form X")
```

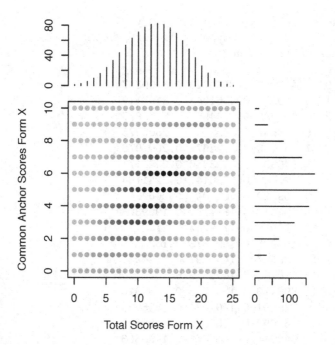

FIGURE 10.4
Smoothed (loglinear) plot of the common anchor scores against total scores
on form X.

There are a variety of presmoothing options available in **equate**. One
option involves fitting a log linear model. Log linear presmoothing can be
specified for forms X and Y as follows:

```
smooth_x <- presmoothing(negd_x, smoothmethod = "loglinear")
smooth_y <- presmoothing(negd_y, smoothmethod = "loglinear")
```

Figure 10.4 shows the smoothed distribution of form X. The irregularities
have been removed from the marginal distributions. The decision to pres-
mooth should be based on the extent of the irregularities and the equating
method considered. In particular, when irregularities are great and equipercentile equating is used, then presmoothing should be used.

```
plot(smooth_x, xlab = "Total Scores Form X",
    ylab = "Common Anchor Scores Form X")
```

Because we are using linear equating and not equipercentile equating, instead of using the presmoothed distributions of form X and Y, we will use the
untransformed distributions and specify the Tucker equating method. We can
extract the concordance table using $concordance.

```
negd_tucker <- equate(negd_x, negd_y,
                      type = "linear", method = "tucker")
negd_tucker$concordance
   scale        yx       se.n       se.g
1      0  4.439083  0.3480743  0.3400317
2      1  5.220574  0.3261043  0.3197119
3      2  6.002065  0.3044597  0.2996877
4      3  6.783557  0.2832149  0.2800225
5      4  7.565048  0.2624671  0.2607975
6      5  8.346539  0.2423440  0.2421177
7      6  9.128031  0.2230147  0.2241194
8      7  9.909522  0.2047041  0.2069804
9      8 10.691013  0.1877108  0.1909323
10     9 11.472505  0.1724244  0.1762733
11    10 12.253996  0.1593373  0.1633778
12    11 13.035487  0.1490297  0.1526932
13    12 13.816978  0.1421079  0.1447100
14    13 14.598470  0.1390783  0.1398916
15    14 15.379961  0.1401934  0.1385683
16    15 16.161452  0.1453580  0.1408389
17    16 16.942944  0.1541655  0.1465363
18    17 17.724435  0.1660372  0.1552838
19    18 18.505926  0.1803692  0.1666016
20    19 19.287418  0.1966241  0.1800056
21    20 20.068909  0.2143650  0.1950662
22    21 20.850400  0.2332530  0.2114297
23    22 21.631892  0.2530314  0.2288166
24    23 22.413383  0.2735071  0.2470111
25    24 23.194874  0.2945347  0.2658473
26    25 23.976366  0.3160041  0.2851982
```

The output shows the concordance table with the original scores from form X, the corresponding scores on the scale of form Y, and the estimated standard errors. The results indicate that the adjustment of the scores from form X is not linear. The gap between the original scores and the adjusted scores is larger on the tails of the score scale (e.g., X scores ranging from 0 to 4 and X scores ranging from 23 to 25). Also, the standard errors for the adjusted X scores (labeled as yx) are higher on the low and high ends of the score distribution.

10.4   Summary

This chapter briefly introduced observed score equating using the **equate** package. There were several examples shown for equivalent groups and an example using a nonequivalent group design. In addition, the presmoothing procedure was briefly presented with an example. For more information about the other equating methods and functions available in the **equate** package, we encourage our readers to check out Albano (2016). IRT-based equating and linking, not discussed in this chapter, can be performed using the **plink** package (Weeks, 2010).

11

Measurement Invariance and Differential Item Functioning

11.1 Chapter Overview

In this chapter, we focus on two frameworks for detecting measurement bias. First, we begin with a brief description of measurement invariance as a factorial approach for detecting scale-level (i.e., overall) bias, and demonstrate how to assess measurement invariance using the **lavaan** package (Rosseel, 2012). Then, we introduce differential item functioning (DIF) as an indicator of item-level bias, describe uniform and nonuniform DIF, and then demonstrate how to detect DIF with the Mantel-Haenszel (MH) method, logistic regression, and the item response theory likelihood ratio (IRT-LR) test using the **difR** (Magis, Beland, Tuerlinckx, & De Boeck, 2010) and **mirt** (Chalmers, 2012) packages in R.

11.2 Measurement Invariance

When we intend to use scores from a measurement instrument to make comparisons among different groups of individuals, we need to show that our latent variable(s) are functioning similarly across the subgroups in the sample (e.g., male and female participants). This procedure is referred to as *measurement invariance*. For example, assume that we are developing an instrument to measure individuals' spirituality in daily life. Once we administer this instrument to a sample of participants, we want to make comparisons between male and female participants about their spirituality levels. If some of the items in the instrument or the entire instrument do not function similarly for the male and female participants, then our latent variable (i.e., spirituality in daily life) may not have the same meaning across the gender groups. When this issue occurs, it makes no sense to state, for example, that one group has a higher level of spirituality than another group, because the comparison is not based on the same latent variable.

Another situation where measurement invariance is essential occurs when we want to compare scores from an instrument at time 1 to scores from the same instrument at time 2. To be able to make any comparisons between these two time points, we need to ensure the stability of the instrument over time. Continuing from the same spirituality example, if we intend to compare the scores from two administrations of the spirituality instrument (e.g., 2002 and 2017), we need to demonstrate measurement invariance of the instrument over time. If the operational definition of spirituality (i.e., how spiritually is viewed and defined by people) has changed from 2002 to 2017, then we may not have comparable scores from the two administrations, despite using the same instrument.

We often implicitly assume measurement invariance when we make comparisons on some construct across groups. For example, we assume that the construct functions similarly when we decide to do regressions and t-tests. Latent variable modeling provides a means for rigorously assessing this assumption.

11.2.1   Assessing Measurement Invariance

In the following example, we use the `interest` data set that we first introduced in Chapter 4. This is a fabricated data set that includes measures of cognitive, personality, and vocational interest. Using the cognitive measures in the `interest` data set, we assess measurement invariance across gender. Before we begin the analysis, we first load the **lavaan** and **hemp** packages using the `library` command.

```
library("hemp")
library("lavaan")
```

In the `interest` data set, the gender variable is coded such that 1 is male and 2 is female. To make our analysis easier to follow, we recode the gender variable into a qualitative variable using the `ifelse` function. The new gender variable has the labels "male" and "female" for the participants instead of 1 and 2.

```
interest$gender_f <- ifelse(interest$gender == 1,
                            "male", "female")
```

Recall from Chapter 4 that we identified two cognitive factors using exploratory factor analysis: a verbal factor and a mathematical factor. We define this factor structure using the **lavaan** syntax and save it to `cog_mod`. Then we run a confirmatory factor analysis (CFA) using the `cfa` function, save the results as `cog_fit`, and finally print some fit indices for the estimated model using the `fitMeasures` function.

```
cog_mod <- 'verb  =~ vocab + reading + sentcomp
            math =~ mathmtcs + geometry + analyrea'

cog_fit <- cfa(cog_mod, data = interest)
fitMeasures(cog_fit, fit.measures = c("cfi", "tli", "rmsea"))
```

```
  cfi   tli rmsea
0.996 0.992 0.051
```

Our model has good fit based on the CFI, TLI, and RMSEA. If the model did not fit the data well, then it would make no sense to consider measurement invariance. Because our model fits the data well, we can examine measurement invariance by gender.

11.2.1.1   Configural Invariance

The first step in demonstrating measurement invariance is to show that the model fits for each of our groups. This can be done in two ways: (1) we can fit the data to each subset of the data; (2) we can use a multiple-group CFA model. We will use the second approach and do this in **lavaan** by simply adding the **group = "gender_f"** argument when we use the **cfa** function. This argument allows all of the parameters to be freely estimated across the groups with no equality constraints. Therefore, the factor loadings, intercepts, and variances are not equal across the two models (the male and female models). Note that we should not use standardized coefficients here, as we have no reason to expect the variances to be the same across the groups *a priori*.

```
configural <- cfa(cog_mod, data = interest, group = "gender_f")
fitMeasures(configural, fit.measures = c("cfi", "tli", "rmsea"))
```

```
  cfi   tli rmsea
0.998 0.996 0.035
```

The model fits the data quite well for each group based on the fit indices. Because the configural model fits well, we can now start constraining the parameters to be equal across the groups and more thoroughly examine measurement invariance by gender.

11.2.1.2   Weak Invariance

To assess weak (metric) invariance, we constrain the factor loadings to be equal across the groups. This constraint indicates that the factor(s) have the same meaning across the groups. In other words, it addresses the question of whether the groups are attributing the same meaning to the latent constructs. If the weak invariance assumption does not hold, then the meaning of the manifest variables would be different across the groups. At this stage, and in any future steps, if we have only partial invariance (which we can assess), then we will constrain the invariant parameters to be equal and allow the other ones to be freed across the groups.

To fit the weak invariance model in **lavaan**, we need to pass the group.equal = "loadings" argument to the cfa function. This argument ensures that the factor loadings are the same between the male and female participants in the interest data set.

```
weak_invariance <- cfa(cog_mod, data = interest,
                    group = "gender_f",
                    group.equal = "loadings")
fitMeasures(weak_invariance,
          fit.measures = c("cfi", "tli", "rmsea"))
```

```
  cfi    tli rmsea
1.000 1.000 0.004
```

Based on the fit indices from weak_invariance, we have evidence supporting the weak invariance assumption for our model. Because all of the measurement invariance models are nested within one another, we can formally assess whether the less constrained model (the configural model) fits significantly better than the more restrictive weak invariant model using a chi-square test of difference. Because our goal is to show measurement invariance, we do not want to reject the null hypothesis that the configural model is not an improvement in model fit over the weak invariance model.

```
anova(weak_invariance, configural)
```

```
Chi Square Difference Test
```

	Df	AIC	BIC	Chisq	Chisq diff	Df diff
configural	16	3081.8	3215.6	18.413		
weak_invariance	20	3075.4	3195.2	20.048	1.6348	4

	Pr(>Chisq)
configural	
weak_invariance	0.8025

Based on the output above, we fail to reject the null hypothesis. This is evidence supporting weak invariance. Based on this result, we can conclude that we have weak invariance and move on to assessing strong invariance.

11.2.1.3 Strong Invariance

Strong (scalar) invariance adds the additional constraint that the intercepts are also equal across the groups. In the strong invariance model, both loadings and intercepts are set to be equal across the groups. This implies that the meaning of the construct (the factor loadings) and the levels of the underlying manifest variables (intercepts) are equal in both groups. If we have strong invariance, then we are able to compare differences in the means across the latent construct(s). It is at this stage in assessing measurement invariance that we can estimate and examine factor means.

For testing strong invariance, we update the `group.equal` argument as `group.equal = c( "loadings", "intercepts")` in the model, save the results of the model as `strong_invariance`, and then print the model fit indices for the estimated model.

```
strong_invariance <- cfa(cog_mod, data = interest,
              group = "gender_f",
              group.equal = c( "loadings", "intercepts"))
fitMeasures(strong_invariance,
          fit.measures = c("cfi", "tli", "rmsea"))
```

```
  cfi   tli rmsea
1.000 1.001 0.000
```

The output shows that our fit indices continue to improve. Based on the fit indices alone, we appear to have met the assumption of strong invariance. However, we should also formally assess this assumption using the `anova` function. This time, our constrained model is the `strong_invariance` model and the less constrained model is the `weak_invariance` model.

```
anova(strong_invariance, weak_invariance)
```

```
Chi Square Difference Test

                  Df    AIC    BIC  Chisq Chisq diff
weak_invariance   20 3075.4 3195.2 20.048
strong_invariance 24 3070.6 3176.3 23.230     3.1816
                  Df diff Pr(>Chisq)
weak_invariance
strong_invariance      4     0.5279
```

The results show that we again fail to reject the null hypothesis. We could stop and examine the factor means at this point; however, it would be more beneficial to demonstrate strict invariance prior to examining model output.

11.2.1.4   Strict Invariance

For strict invariance, we add the additional constraint of equal residual variances for the manifest variables across the groups. If we want to make comparisons based on the raw scores (e.g., based on the sum or mean of the scores on the indicators of a latent variable), we need to make sure that our model meets the strict invariance assumption. This is because the observed variance is the sum of true score variance and residual/error variance. If the residual variances are the same, then they have the same amount of true score variance. As with strong invariance, we can estimate and compare the latent means.

Similar to the previous model syntax, we need to update the `group.equal` argument for this model. This time, we use `group.equal = c( "loadings", "intercepts", "residuals")` to constrain all of these parameters to be the same across gender. We save the results as `strict_invariance`.

```
strict_invariance <- cfa(cog_mod, data = interest,
        group = "gender_f",
        group.equal = c( "loadings", "intercepts", "residuals"))
fitMeasures(strict_invariance,
        fit.measures = c("cfi", "tli", "rmsea"))
```

```
  cfi   tli rmsea
1.000 1.002 0.000
```

The fit indices from `strict_invariance` again look quite good. Next, we formally assess whether this model is a significant improvement in fit. For this test, we compare `strict_invariance` to `strong_invariance`.

```
anova(strict_invariance, strong_invariance)
```

```
Chi Square Difference Test
```

	Df	AIC	BIC	Chisq	Chisq diff
strong_invariance	24	3070.6	3176.3	23.230	
strict_invariance	30	3062.7	3147.2	27.264	4.0347

	Df diff	Pr(>Chisq)
strong_invariance		
strict_invariance	6	0.672

The results show that we fail to reject the null hypothesis for this test and conclude that our model meets the strict invariance assumption. We want to note that meeting the strict invariance assumption is extremely rare in practice and that strong invariance is usually sufficient to compare factor means.

Based on the results for strict invariance, we could safely compare factors based on the raw sum or mean scores as they are measured with equal reliability across the groups. However, since we have already created factors, we examine the factor means, which we can get from the summary function.

```
summary(strict_invariance)
```

lavaan (0.5-20) converged normally after 42 iterations

```
Number of observations per group
female                                          122
male                                            128

Estimator                                        ML
Minimum Function Test Statistic              27.264
Degrees of freedom                               30
P-value (Chi-square)                          0.609
```

Chi-square for each group:

```
female                                       15.063
male                                         12.201
```

Group 1 [female]:

Latent Variables:

	Estimate	Std.Err	Z-value	P(>|z|)
verb =~				
vocab	1.000			
reading (.p2.)	0.898	0.044	20.500	0.000
sentcmp (.p3.)	0.904	0.044	20.743	0.000

Group 2 [male]:

Latent Variables:

	Estimate	Std.Err	Z-value	P(>|z|)
verb =~				
vocab	1.000			
reading (.p2.)	0.898	0.044	20.500	0.000
sentcmp (.p3.)	0.904	0.044	20.743	0.000

Intercepts:

	Estimate	Std.Err	Z-value	P(>|z|)
vocab (.16.)	0.169	0.093	1.813	0.070
reading (.17.)	0.206	0.088	2.343	0.019
sentcmp (.18.)	0.145	0.088	1.642	0.101
mthmtcs (.19.)	-0.048	0.097	-0.489	0.625
geomtry (.20.)	-0.019	0.089	-0.208	0.835

analyre (.21.)	0.032	0.094	0.345	0.730
verb	-0.154	0.123	-1.247	0.212
math	0.299	0.129	2.320	0.020

We have abridged much of the output from the summary function to focus on the new information provided when we examine measurement invariance. For a more complete description of the output from the cfa function, our readers can review Chapter 4. At the top of the output, we see the number of observations per group, followed by the chi-square of the present model, and the contribution of each group towards the chi-square statistic. Because the groups are independent, the sum of each group's contribution is the total chi-square statistic. Then we have the estimated parameters first printed for the baseline group, females in this case, followed by the comparison group, males. The factor loadings for the mathematics factor and the covariances and variances have been removed from the output. Focusing on the factor loadings for reading, we note there is a (.p2.) next to it. This is how **lavaan** handles constraints, by labeling the parameters. In this example, the name .p2. corresponds to the factor loading of reading onto verbal for both females and males. This is also the case with the other parameters (e.g., .p3. for sentcmp in the same output).

At the bottom of the intercepts table, we see the means for the verbal and mathematics factors for the male group. A statistically significant intercept means that there are significant group differences on the factor. We find that there are no statistically significant differences between the groups on the verbal factor, but we find that the male group is expected to be 0.299 higher on the math factor than the female group.

Overall, we conclude that we have strict invariance, that there were no differences on the verbal factor between groups, and that males are expected to be higher than females on the mathematics factor.

11.2.1.5   Assessing Partial Invariance

In the event that we did not have full invariance at any stage (i.e., weak, strong, and strict invariance), then we could focus on alternative solutions based on partial invariance. We can examine partial invariance with the lavTestScore function in **lavaan**. This function allows us to see the effect of releasing an equality constraint across the groups. This is opposite to the modindices function, which only shows modification indices for newly added parameters associated with new paths. In the context of partial invariance, the parameters are not newly estimated in the model, but instead, we just let the model freely estimate them across groups, and that is why we use lavTestScore and not modindices.

In the following example, we examine partial invariance for our strong invariance model (i.e., `strong_invariance`) using the `lavTestScore` function.

```
lavTestScore(strong_invariance)
```

```
$test

total score test:

    test    X2 df p.value
1 score  4.822 10   0.903

$uni

univariate score tests:

       lhs op   rhs    X2 df p.value
1    .p2. ==  .p25. 0.520  1   0.471
2    .p3. ==  .p26. 0.669  1   0.414
3    .p5. ==  .p28. 0.082  1   0.775
4    .p6. ==  .p29. 0.000  1   0.997
5   .p16. ==  .p39. 0.533  1   0.465
6   .p17. ==  .p40. 0.247  1   0.619
7   .p18. ==  .p41. 0.159  1   0.690
8   .p19. ==  .p42. 2.540  1   0.111
9   .p20. ==  .p43. 0.311  1   0.577
10  .p21. ==  .p44. 1.765  1   0.184
```

The first part of the output is a multivariate score test (a Lagrange multiplier test), which is a test of the null hypothesis that there would be no model improvement if we released all the constraints together. The results show that we fail to reject this null hypothesis. The first, second, and third columns correspond to the parameters that are being considered for unconstrained estimation (i.e., being freely estimated across the groups). The fourth column is the univariate score tests (i.e., the chi-square difference tests). Again note that 3.84 is the critical value for 1 degree of freedom. Unsurprisingly because we found strong invariance, we find that freeing any equality would not result in a significant statistical improvement in model fit. However, had this not been the case, we might expect to see a large chi-square value in the X2 column. The largest X2 value is associated with .p19. == .p42.. In order to see what parameter this refers to, we need to run the `parTable` function and look up the entry for .p19. in the table.

```
parTable(strong_invariance)
   id      lhs op      rhs user group free ustart exo label
19 19 mathmtcs ~1            0     1   17     NA   0 .p19.
```

We find that this corresponds to mathmtcs $\sim$ 1, which indicates that we should free this intercept across the groups. If we want to include this in a call to `cfa`, we would copy and paste mathmtcs $\sim$ 1 into the `group.partial` argument. The following code shows how we can free the intercept parameter.

```
strong_invariance_p <- cfa(cog_mod, data = interest,
                           group = "gender_f",
                           group.equal = c( "loadings", "intercepts"),
                           group.partial = "mathmtcs ~ 1")
```

11.3   Differential Item Functioning

Differential item functioning (DIF) is used in educational and psychological assessments to detect item-level bias. DIF occurs due to a conditional dependency between group membership of examinees (e.g., male versus female) and item performance after controlling for the latent trait. If the item is dichotomously scored, item performance refers to the probability of answering the item correctly; if the item is polytomously scored, then item performance refers to the probability of selecting or endorsing a particular response category over the other (ordered or non-ordered) response categories.

As a result of DIF, a biased item provides either a constant advantage for a particular group (uniform DIF) or an advantage varying in magnitude and/or in direction across the latent trait continuum (nonuniform DIF). In the context of DIF, the focal group refers to the examinees who are suspected to be at a disadvantage compared to the examinees from the reference group when responding to the items on the test. If an item exhibits uniform DIF, then the performance of the focal group tends to be constantly worse than the performance of the reference group along the latent trait continuum. After controlling for the latent trait levels, the focal group has a lower probability of answering the item correctly than the reference group. If the item exhibits nonuniform DIF, the difference between the focal and reference groups does not have a single direction. That is, although one group may outperform another group, this relationship is reversed after a certain point on the latent trait continuum.

Figure 11.1 demonstrates an example of uniform and nonuniform DIF for a dichotomous item. In each plot, the item characteristic curves (ICCs; see Chapter 5 for more information on ICCs) for the focal and reference groups are shown separately. When uniform DIF is present, the focal group is less likely to answer the item correctly compared to the reference group. When nonuniform

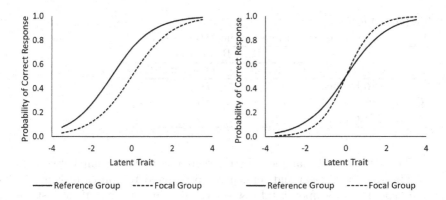

FIGURE 11.1
Uniform (left) and nonuniform (right) DIF in a dichotomous item.

DIF is present, the ICCs for the reference and focal groups intersect at $\theta = 0$ on the latent trait continuum. This suggests that while the reference group has a higher probability of answering the item correctly than the focal group until $\theta = 0$, the item begins to favor the focal group over the reference group after $\theta = 0$.

There are many statistical methods available for detecting uniform and nonuniform DIF in dichotomously and polytomously scored items. In this chapter, we focus on three methods, namely the Mantel-Haenszel (MH) method (Mantel & Haenszel, 1959), logistic regression (Swaminathan & Rogers, 1990), and the item response theory likelihood ratio (IRT-LR) test (Thissen, Steinberg, & Wainer, 1993). The IRT-LR test is considered a parametric DIF detection method because it requires using a particular IRT model for estimating item parameters and latent trait levels, which are then used during the course of DIF analysis. Unlike the IRT-LR test, the MH and logistic regression methods are considered nonparametric DIF detection methods because these methods only involve raw item and test scores in detecting DIF (i.e., IRT is not required). In the following sections, we briefly describe each DIF detection method and demonstrate its implementation in R.

11.3.1   The Mantel-Haenszel (MH) Method

The MH method (Mantel & Haenszel, 1959) is one of the most widely used DIF detection methods because it is easy to compute and interpret compared to other DIF detection methods. The MH method is capable of detecting only uniform DIF. Therefore, when nonuniform DIF is present, the MH method may not be an appropriate approach.

TABLE 11.1

A $2 \times 3 \times 2$ contingency table for the MH method.

		Item Responses		
		1=Correct	0=Incorrect	Total
	Reference	A_k	B_k	n_{Rk}
Group	Focal	C_k	D_k	n_{Fk}
	Total	n_{1k}	n_{0k}	T_k

The MH method examines whether there is any relationship between group membership (i.e., focal and reference groups) and responses to a particular item, conditionally upon total raw scores. To compute the MH statistic for a given item, a $2 \times 3 \times 2$ contingency table is created with group membership and type of response (correct or incorrect) as rows and columns. Table 11.1 shows the elements of the contingency table required for computing the MH statistic. Using k as an index of examinee groups separated based on their total raw scores ($k = 1, 2, 3, \ldots, m$), A_k and C_k are the numbers of examinees in the reference and focal groups who answered the item correctly, B_k and D_k are the numbers of examinees in the reference and focal groups who answered the item incorrectly, n_{1k} and n_{0k} are the total numbers of examinees who answered the item correctly and incorrectly, n_{Rk} and n_{Fk} are the total numbers of examinees in the reference and focal groups, and T_k is the total number of examinees in group k.

Using the frequencies listed in Table 11.1, the MH statistic ($\hat{\alpha}_{MH}$) can be computed as follows:

$$\hat{\alpha}_{MH} = \frac{\sum_{k=1}^{m} \left[\frac{B_k C_k}{T_k} \frac{A_k D_k}{B_k C_k} \right]}{\sum_{k=1}^{m} \frac{B_k C_k}{T_k}} = \frac{\sum_{k=1}^{m} \frac{A_k D_k}{T_k}}{\sum_{k=1}^{m} \frac{B_k C_k}{T_k}}. \tag{11.1}$$

$\hat{\alpha}_{MH} > 1$ indicates significant DIF favoring the reference group whereas $\hat{\alpha}_{MH} < 1$ indicates significant DIF favoring the focal group. The $\hat{\alpha}_{MH}$ statistic is often transformed into $\hat{\lambda}_{MH}$, which is the natural logarithm of the $\hat{\alpha}_{MH}$ statistic, $\hat{\lambda}_{MH} = ln(\hat{\alpha}_{MH})$. This alternative statistic, λ, is a symmetric measure of DIF, where $\lambda = 0$ indicates no DIF, $\lambda > 0$ indicates DIF favoring the reference group, and $\lambda < 0$ indicates DIF favoring the focal group. The more $\hat{\lambda}_{MH}$ deviates from zero, the more significant DIF becomes.

To test whether DIF is statistically significant, the Mantel-Haenszel chi-square test can be used (Mantel & Haenszel, 1959). The Mantel-Haenszel chi-square statistic (χ^2_{MH}) can be computed as follows (Penfield & Camilli, 2006):

$$\chi^2_{MH} = \frac{\left[|\sum_{k=1}^{m}[A_k - E(A_k)]| - 0.5\right]^2}{\sum_{k=1}^{m} V(A_k)}, \tag{11.2}$$

where

$$E(A_k) = \frac{n_{Rk}n_{1k}}{T_k}, \text{ and} \tag{11.3}$$

$$V(A_k) = \frac{n_{Rk}n_{Fk}n_{1k}n_{0k}}{T_k^2(T_k - 1)}. \tag{11.4}$$

The χ^2_{MH} statistic is distributed approximately as a chi-square variable with one degree of freedom. If the χ^2_{MH} statistic is statistically significant for a particular item, the item should be flagged for exhibiting uniform DIF. In addition to the chi-square significance test, Holland and Thayer (1988) proposed an alternative index called the MH Delta index using the $\hat{\alpha}_{MH}$ statistic. This index functions like an effect size measure and categorizes the severity of DIF. This procedure is also known as *ETS delta classification*. The MH delta index can be computed as follows:

$$\Delta_{MH} = -2.35\hat{\alpha}_{MH}, \tag{11.5}$$

where Δ_{MH} is negative for DIF favoring the reference group and Δ_{MH} is positive for DIF favoring the focal group. In addition, $|\Delta_{MH}| \leq 0$ corresponds to "A: Negligible DIF", $1 < |\Delta_{MH}| \leq 1.5$ corresponds to "B: Moderate DIF", and $|\Delta_{MH}| > 1.5$ corresponds to "C: Large DIF" (Holland & Thayer, 1988).

In the following example, we use the VerbAggWide data set from the **hemp** package and the **difR** package (Magis et al., 2010) to demonstrate the MH method in detecting uniform DIF. As we described in Chapter 6, the VerbAggWide data set contains the responses of 316 participants to a series of questions related to verbal aggression. The first three variables in the data set are id (participant identification number), Anger (participants' anger scores), and Gender (participants' gender, where M refers to male and F refers to female). The remaining variables (e.g., S1WantCurse, S1WantScold, and S1WantShout) are the responses to the questions on verbal aggression.

Before we begin the DIF analysis, we first create a new data set called VerbAgg, which only includes Gender and the questions in the VerbAggWide data set. Then we recode the original responses (0: No, 1: Perhaps, and 2: Yes) in VerbAgg into dichotomous responses (0: No, 1: Perhaps or Yes) because the

difR package requires the responses to be dichotomous. We use the `apply` and `ifelse` functions to recode the questions (i.e., columns 2 to 25).

```
VerbAgg <- VerbAggWide[, c(3, 4:27)]
VerbAgg[, 2:25] <- apply(VerbAgg[, 2:25], 2,
                         function(x) ifelse(x == 0, 0, 1))
```

Next, we use the `difMH` function in the **difR** package to conduct DIF analysis with the MH method. In the `difMH` function, we specify the data set to be used (`VerbAgg`), the variable representing the grouping variable (`Gender`), and the value that designates the focal group (i.e., F for the female participants). We save the results as `results_MH`, and then print them.

```
library("difR")
results_MH <- difMH(Data = VerbAgg, group = "Gender",
                    focal.name = "F")
results_MH
```

```
Mantel-Haenszel Chi-square statistic:

Stat.  P-value
S1WantCurse 1.7076 0.1913
S1WantScold 2.1486 0.1427
S1WantShout 0.9926 0.3191
S2WantCurse 1.9302 0.1647
S2WantScold 2.9540 0.0857  .
S2WantShout 9.6032 0.0019  **
S3WantCurse 0.0013 0.9711
S3WantScold 0.6752 0.4112
S3WantShout 0.8185 0.3656
S4wantCurse 1.6292 0.2018
S4WantScold 0.0152 0.9020
S4WantShout 4.1188 0.0424  *
S1DoCurse   0.1324 0.7160
S1DoScold   2.7501 0.0972  .
S1DoShout   0.0683 0.7938
S2DoCurse   6.3029 0.0121  *
S2DoScold   6.8395 0.0089  **
S2DoShout   0.2170 0.6414
S3DoCurse   5.7817 0.0162  *
S3DoScold   3.8880 0.0486  *
S3DoShout   0.2989 0.5846
S4DoCurse   1.1220 0.2895
S4DoScold   1.4491 0.2287
S4DoShout   0.8390 0.3597

Signif. codes: 0 '***' 0.001 '**' 0.01 '*' 0.05 '.' 0.1 ' ' 1
```

```
Detection threshold: 3.8415 (significance level: 0.05)

Items detected as DIF items:

S2WantShout
S4WantShout
S2DoCurse
S2DoScold
S3DoCurse
S3DoScold
```

The output consists of two sections. The first part of the output shows the MH chi-square statistics for the items and their corresponding p values. The results from the MH chi-square test indicate that six items in the `VerbAgg` data set have been flagged at the alpha level of $\alpha = .05$ for showing uniform DIF between the male and female participants. These items are listed at the bottom of the output. The second part of the output shows the effect size using ETS' delta classification for the items. The results show that nine items have been flagged for showing "C: Large DIF," seven items haven flagged for showing "B: Moderate DIF," and the remaining items have been classified as "A: Negligible DIF."

```
Effect size (ETS Delta scale):

Effect size code:
'A': negligible effect
'B': moderate effect
'C': large effect

alphaMH deltaMH
S1WantCurse  0.5881   1.2476 B
S1WantScold  0.5649   1.3420 B
S1WantShout  0.6906   0.8701 A
S2WantCurse  0.5156   1.5567 C
S2WantScold  0.5051   1.6052 C
S2WantShout  0.3472   2.4861 C
S3WantCurse  1.0595  -0.1358 A
S3WantScold  1.3901  -0.7741 A
S3WantShout  0.6544   0.9965 A
S4wantCurse  0.5935   1.2260 B
S4WantScold  0.9173   0.2028 A
S4WantShout  0.4263   2.0036 C
S1DoCurse    1.2551  -0.5340 A
S1DoScold    2.0021  -1.6313 C
S1DoShout    0.8499   0.3821 A
```

```
S2DoCurse    3.1159 -2.6709 C
S2DoScold    2.6693 -2.3072 C
S2DoShout    1.2608 -0.5447 A
S3DoCurse    2.1662 -1.8165 C
S3DoScold    2.1153 -1.7606 C
S3DoShout    1.5690 -1.0585 B
S4DoCurse    1.5518 -1.0327 B
S4DoScold    1.5661 -1.0541 B
S4DoShout    0.6229  1.1123 B


Effect size codes: 0 'A' 1.0 'B' 1.5 'C'
(for absolute values of 'deltaMH')
```

The saved output from the difMH function can also be used to examine DIF graphically (Figure 11.2). Using the plot function, we can see the items that have been flagged based on the MH chi-square test. The items flagged for DIF appear above the horizontal line for the threshold chi-square value ($\chi^2 = 3.84$ for the significance level of $\alpha = .05$). The alpha level can be modified by setting the alpha option in the difMH function (e.g., alpha = .01).

```
plot(results_MH)
```

The difMH function includes other useful options, such as anchor to specify anchor (DIF-free) items, purify for an iterative process to purify anchor items, and p.adjust.method for a p-value adjustment in multiple comparisons (e.g., the Benjamini-Hochberg adjustment). For example, we could run the same DIF analysis with purification and p-value adjustment.

```
results_MH <- difMH(Data = VerbAgg, group = "Gender",
                    focal.name = "F", purify = TRUE,
                    p.adjust.method = "BH")
```

11.3.2   Logistic Regression

The logistic regression approach (Swaminathan & Rogers, 1990) has been one of the most popular DIF detection methods because of its computational simplicity and its capability to detect both uniform and nonuniform DIF. This DIF detection method is based on the comparison of a series of logistic regression models where the probability of answering a dichotomous item correctly (or endorsing a particular response option) is predicted by examinees' trait estimates (X), group membership (G), and the interaction of examinees' trait estimates with group membership (GX). To explain the logistic regression approach for detecting DIF, consider the following logistic regression model:

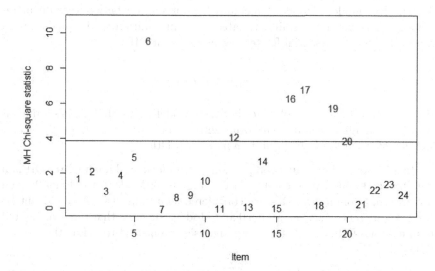

FIGURE 11.2
Flagged items based on the MH chi-square test.

$$P(Y = 1|z) = \frac{e^z}{1 + e^z}, \tag{11.6}$$

where $P(Y = 1|z)$ is the probability of giving a correct response to an item under investigation for DIF and z is a linear equation based on one or more predictors.

In the logistic regression approach, the baseline model (Model 0) only includes examinees' trait estimates (X) as a single predictor

$$z = \beta_0 + \beta_1 X, \tag{11.7}$$

where X is examinees' trait estimates (e.g., total raw scores). The next model (Model 1) includes group membership (G) as a predictor in addition to examinees' trait estimates, which is essential for testing uniform DIF

$$z = \beta_0 + \beta_1 X + \beta_2 G, \tag{11.8}$$

where $G = 1$ for the focal group and $G = 0$ for the reference group. If the pseudo R-squared difference between Model 0 and Model 1 is significant or β_2 is significantly different from zero (i.e., $\beta_2 \neq 0$), then the item exhibits uniform DIF.

The last model (Model 2) includes the interaction between group membership and examinees' trait estimates (GX) in addition to the predictors in Model 1, which is essential for testing nonuniform DIF.

$$z = \beta_0 + \beta_1 X + \beta_2 G + \beta_3 GX. \tag{11.9}$$

If the pseudo R-squared difference between Model 1 and Model 2 is significant (or β_3 is significantly different from zero) regardless of whether β_2 is also significant, then the item exhibits nonuniform DIF.

In addition to individual comparisons for uniform and nonuniform DIF, an omnibus test can also be conducted for comparing Model 0 to Model 2 when uniform and nonuniform DIF are considered simultaneously. A significant R-squared difference between Model 0 and Model 2 suggests that the investigated item exhibits DIF, and so follow-up analyses are needed to define the type of DIF.

To demonstrate the detection of DIF using the logistic regression approach, we use the `difLogistic` function in the **difR** package in the following example. As for the MH method, we use `VerbAgg` for demonstrating the logistic regression approach. In the `difLogistic` function, we first specify the data set to be used (`VerbAgg`), the variable representing the group variable (`Gender`), the value that designates the focal group (`focal.name = "F"` to set the female participants as the focal group), and the type of test (`type = "both"` for simultaneous test of uniform and nonuniform DIF). We save the results as `logistic_fit` and print them at the end.

```
logistic_fit <- difLogistic(Data = VerbAgg, group = "Gender",
                            focal.name = "F", type = "both")
logistic_fit
```

```
Logistic regression DIF statistic:

            Stat.    P-value
S1WantCurse  2.0014   0.3676
S1WantScold  3.3541   0.1869
S1WantShout  2.4742   0.2902
S2WantCurse  4.7296   0.0940 .
S2WantScold  4.1404   0.1262
S2WantShout 11.4111   0.0033 **
S3WantCurse  1.6061   0.4480
S3WantScold  1.6331   0.4419
S3WantShout  2.6989   0.2594
S4wantCurse  2.4547   0.2931
S4WantScold  2.0997   0.3500
S4WantShout  3.6877   0.1582
```

```
S1DoCurse      1.2196   0.5435
S1DoScold      4.7304   0.0939 .
S1DoShout      1.0456   0.5929
S2DoCurse      7.6935   0.0213 *
S2DoScold     10.2622   0.0059 **
S2DoShout      1.7016   0.4271
S3DoCurse      7.2379   0.0268 *
S3DoScold      5.8680   0.0532 .
S3DoShout      1.2763   0.5283
S4DoCurse      2.9521   0.2285
S4DoScold      2.6956   0.2598
S4DoShout      1.3524   0.5085


Signif. codes:  0 '***' 0.001 '**' 0.01 '*' 0.05 '.' 0.1 ' ' 1

Detection threshold: 5.9915 (significance level: 0.05)

Items detected as DIF items:

S2WantShout
S2DoCurse
S2DoScold
S3DoCurse
```

Similar to the output from the `difMH` function, the output from the `difLogistic` function consists of two sections. The top part of the output (shown above) shows the likelihood ratio test statistics obtained by comparing the fit of Model 0 and Model 2 and the corresponding p values. The results show that four items have been flagged for DIF at the significance level of $\alpha = .05$. These items are listed at the end of the first part of the output.

The second part of the output (shown below) shows the effect sizes computed using the pseudo R-squared difference between Model 0 and Model 2. The column ZT refers to the effect size method proposed by Zumbo and Thomas (1997) as $\Delta R^2 \leq .13$ for negligible DIF, $.13 < \Delta R^2 \leq .26$ for moderate DIF, and $\Delta R^2 > .26$ for large DIF. The column JG refers to another effect size method proposed by Jodoin and Gierl (2001) as $\Delta R^2 \leq .035$ for negligible DIF, $.035 < \Delta R^2 \leq .07$ for moderate DIF, and $\Delta R^2 > .07$ for large DIF. The results show that the effect size was "A: Negligible DIF" for all of the items based on both effect size measures.

```
Effect size (Nagelkerke's R^2):

Effect size code:
'A': negligible effect
'B': moderate effect
```

```
'C': large effect

             R^2    ZT JG
S1WantCurse 0.0067 A  A
S1WantScold 0.0101 A  A
S1WantShout 0.0074 A  A
S2WantCurse 0.0177 A  A
S2WantScold 0.0125 A  A
S2WantShout 0.0343 A  A
S3WantCurse 0.0056 A  A
S3WantScold 0.0050 A  A
S3WantShout 0.0106 A  A
S4wantCurse 0.0085 A  A
S4WantScold 0.0060 A  A
S4WantShout 0.0133 A  A
S1DoCurse   0.0039 A  A
S1DoScold   0.0122 A  A
S1DoShout   0.0033 A  A
S2DoCurse   0.0243 A  A
S2DoScold   0.0277 A  A
S2DoShout   0.0058 A  A
S3DoCurse   0.0232 A  A
S3DoScold   0.0211 A  A
S3DoShout   0.0078 A  A
S4DoCurse   0.0092 A  A
S4DoScold   0.0083 A  A
S4DoShout   0.0057 A  A

Effect size codes:
Zumbo & Thomas (ZT): 0 'A' 0.13 'B' 0.26 'C' 1
Jodoin & Gierl (JG): 0 'A' 0.035 'B' 0.07 'C' 1
```

In addition to the omnibus test of uniform and nonuniform DIF, each type of DIF can also be investigated separately. For example, we can use type = "udif" and type = "nudif" to examine the items for uniform and nonuniform DIF separately.

```
difLogistic(Data = VerbAgg, group = "Gender", focal.name = "F",
        type = "udif")
difLogistic(Data = VerbAgg, group = "Gender", focal.name = "F",
        type = "nudif")
```

The other options in the difLogistic function include criterion = "Wald" for using the Wald test instead of the likelihood ratio statistics, match to select another continuous or discrete variable as a matching criterion, purify = TRUE for iterative purification of the anchor items, and anchor to specify a list of anchor items in the data set.

As we have done for the difMH function, we can use the saved output from the difLogistic function to examine the DIF items graphically (Figure 11.3). Using the plot function, we plot the items that have been flagged based on the likelihood ratio statistic. The items flagged for DIF appear above the horizontal line for the threshold likelihood ratio statistic (5.9915 for the significance level of $\alpha = .05$). The alpha level can be modified using the alpha option in the difLogistic function (e.g., alpha = .01).

```
plot(logistic_fit)
```

The plot function also allows creating individual plots for the flagged items based on the likelihood ratio statistic. The plot creates separate ICCs for the reference and focal groups, which enables a visual examination of the type of DIF (i.e., uniform DIF versus nonuniform DIF) and the severity of the DIF. This plot creates two ICCs based on the matching criterion (e.g., total test score when match = "score" is used in the difLogistic function) on the x-axis and the probability of correct answer (i.e., receiving 1 over 0) on the y-axis. These ICCs can indicate whether the type of DIF is either uniform or nonuniform and which one of the two groups (i.e., reference and focal) has the advantage over the other group.

In the following example, we create an individual DIF plot for item 6 in the VerbAgg data set. We use item = 6 to select item 6 and plot = "itemCurve" to obtain ICCs instead of a plot based on the likelihood ratio statistics. Figure 11.4 shows that item 6 exhibits uniform DIF where the female participants (i.e., focal group) are more likely to choose yes or perhaps over no for item 6 (I would want to shout if I missed a train because a clerk gave me faulty information) compared to the male participants. A similar plot can be created for the other flagged items in the VerbAgg data set.

```
plot(logistic_fit, item = 6, plot = "itemCurve")
```

11.3.3   Item Response Theory Likelihood Ratio Test

The item response theory likelihood ratio (IRT-LR) test (Thissen et al., 1993) is considered one of the most effective methods for detecting uniform and nonuniform DIF for both dichotomously and polytomously scored items. The IRT-LR test involves the comparison of two nested IRT models with a series of likelihood ratio tests. First, a full model, where all item parameters are allowed to vary between the reference and focal groups, is estimated. Next, a restricted model, where the parameters of the item suspected of DIF are constrained to be equal between the reference and focal groups, is estimated.

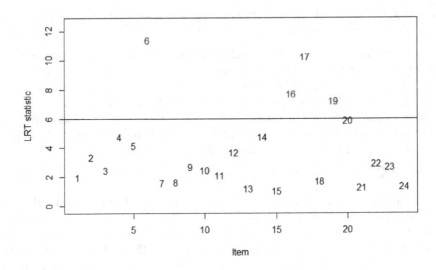

FIGURE 11.3
Flagged items based on the logistic regression approach.

To test whether the item exhibits DIF, a likelihood ratio statistic is computed using the following formula:

$$\text{Likelihood ratio statistic} = -2\ln L_R - (-2\ln L_F), \qquad (11.10)$$

where L_R is the log-likelihood of the restricted model and L_F is the log-likelihood of the full model.

The likelihood ratio statistic follows a chi-square distribution with degrees of freedom equal to the difference in the number of parameters estimated between the restricted and full models. A statistically significant likelihood ratio statistic indicates that an item exhibits DIF, and thus follow-up tests should be performed to identify the type of DIF. For example, only the discrimination parameter of the item is constrained to be equal between the reference and focal groups while the difficulty parameter of the item is estimated for the two groups separately. A significant likelihood ratio statistic from the comparison of this model against the full model described earlier would suggest that the item exhibits nonuniform DIF. A similar test can be conducted by constraining the difficulty parameter of the item to be equal between the two groups in order to identify if the item exhibits uniform DIF.

In the following example, we use the DIF function from the **mirt** package (Chalmers, 2012) to demonstrate how to perform the IRT-LR test to detect uniform and nonuniform DIF. To demonstrate how to detect both uniform and

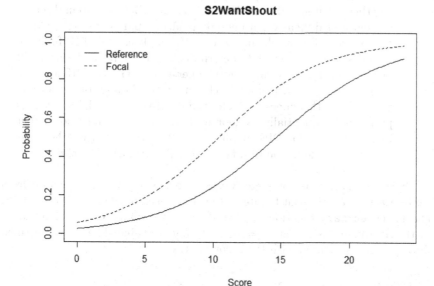

FIGURE 11.4
The item characteristic curves for item 6 (S2WantShout) in the `VerbAgg` data set.

nonuniform DIF, we assume that the items in the `VerbAgg` data set follow the two-parameter IRT model (see Chapter 5 for more information about unidimensional IRT models). The detection of DIF with the `DIF` function involves two steps:

1. The `multipleGroup` function is used to estimate the full model where the item parameters are freely estimated for the reference and focal groups, which are the male and female participants in the `VerbAgg` data set. The `multipleGroup` function is very similar to the `mirt` function that we described in detail in Chapter 5. We first create the `twopl_mod` to define the unidimensional test structure based on the twenty-four items in the `VerbAgg` data set. Next, we pass this model to the `multipleGroup` function as `model = twopl_mod`. Unlike the `mirt` function, the `multipleGroup` function requires specifying a group variable, which is Gender in our example (`group = VerbAgg$Gender`). We save the results of this multi-group analysis as `twopl_fit`.

2. The results from `twopl_fit` are then passed to the `DIF` function as `MGmodel = twopl_fit`. Next, we define which parameters need to

be constrained to be equal in the restricted model. To determine whether the item exhibits DIF, we constrain both the item discrimination and difficulty parameters as which.par = c("a1", "d"). Lastly, we define a scheme for the DIF analysis. The DIF function includes several schemes to determine how the DIF analysis should be performed. For example, with scheme = "add", the DIF analysis is performed by constraining each item in the data set one at a time. Similarly, with scheme = "add_sequential", the DIF analysis is performed sequentially by looping over the items until no further item is flagged for DIF. In this example, we define the scheme as scheme = "add". The results are stored as results_irtlr.

Once the analysis is complete, we use the irtlr_summary function from the **hemp** package to print the items that have been flagged for showing DIF. The irtlr_summary function requires defining the alpha level of significance for the chi-square test, where $\alpha = .05$ is the default value. The steps described above are demonstrated with the following R code.

```
library("mirt")
twopl_mod <- "F = 1 - 24"
twopl_fit <- multipleGroup(data = VerbAgg[, 2:25],
                           model = twopl_mod, SE = TRUE,
                           group = VerbAgg$Gender)
results_irtlr <- DIF(MGmodel = twopl_fit,
                     which.par = c("a1", "d"),
                     scheme = "add")
irtlr_summary(results_irtlr, alpha = 0.05)
```

	Item	chi_square	df	p_value
1	S1DoScold	7.326	2	0.026
2	S2DoCurse	10.699	2	0.005
3	S2DoScold	12.820	2	0.002
4	S3DoCurse	10.312	2	0.006
5	S3DoScold	8.333	2	0.016

The output returned from the irtlr_summary function shows that five items in the VerbAgg data set exhibit gender-related DIF. To determine if these items exhibit either uniform or nonuniform DIF, we conduct a follow-up test by constraining only the item difficulty parameter between the focal and reference groups. We also specify which items should be tested for DIF based on the results from the first analysis. Using the which function, we save the column numbers of the flagged items (e.g., the sixth item in VerbAgg[,2:25]) and save this as DIFitems.

```
DIFitems <- which(colnames(VerbAgg[, 2:25]) %in%
              irtlr_summary(results_irtlr, alpha = .05)$Item)
DIFitems
```

```
[1] 14 16 17 19 20
```

Next, we specify the items to be tested for uniform DIF using `items2test`
= `DIFitems` in the DIF function. We also modify the constrained parameter
as `which.par = "d")` so that the difficulty parameter is fixed between the
focal and reference groups for testing uniform DIF.

```
results_irtlr <- DIF(MGmodel = twopl_fit, which.par = "d",
                scheme = "add", items2test = DIFitems)
irtlr_summary(results_irtlr, alpha = .05)
```

```
         Item chi_square df p_value
1 S1DoScold      5.841   1   0.016
2 S2DoCurse      9.324   1   0.002
3 S2DoScold     11.433   1   0.001
4 S3DoCurse      9.892   1   0.002
5 S3DoScold      8.327   1   0.004
```

The output from the second analysis shows that all of the five items from
the first analysis exhibit uniform DIF, as the p-values are less than the significance level that we set for the DIF analysis ($\alpha = .05$). In addition to the
IRT-LR test results, we can examine the DIF items visually using the `plot`
function. For example, item 14 (S1DoScold) and item 16 (S2DoCurse) are
the two items that have been flagged for showing uniform DIF based on the
IRT-LR test. Using the results from `twopl_fit`, we can draw the ICCs for the
male and female participants for each item.

```
plot(twopl_fit, type = "trace", which.items = c(14, 16),
    par.settings = simpleTheme(lty = 1:2, lwd = 2),
    auto.key = list(points = FALSE, lines = TRUE, columns = 2))
```

Figure 11.5 shows that for both items, the male participants are more
likely to choose yes or perhaps over no compared to the female participants.
The direction of the bias is the same along the latent trait continuum (i.e.,
the male participants continue to have a higher likelihood than the female
participants) because the items exhibit uniform DIF.

The DIF function has other options that could be useful under different
settings. For example, `Wald = TRUE` returns the results from a Wald test instead of a likelihood ratio test. Also, `p.adjust` can be used for adjusting
the p-values from the likelihood ratio test. For example, `p.adjust = "BH"`
and `p.adjust = "bonferroni"` would return p-values adjusted based on the
Benjamini-Hochberg method (Benjamini & Hochberg, 1995) and Bonferroni's
correction, respectively. Lastly, the item plots can be obtained directly from
the DIF function by specifying `plotdif = TRUE`.

In sum, the DIF methods that we have discussed in this chapter are the
most widely used methods for detecting uniform and nonuniform DIF. Al-

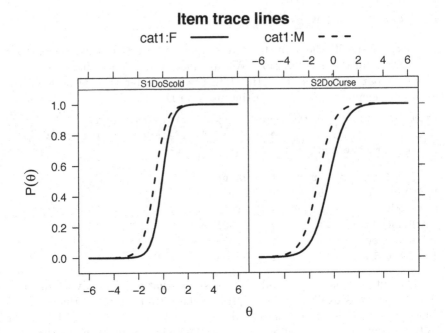

FIGURE 11.5
Item characteristic curves showing uniform DIF in items 14 and 16.

though our examples have focused on dichotomous items in the `VerbAgg` data set, these methods can also be used for polytomous items. In addition, these DIF detection methods can also be used with multidimensional test structures where two or more latent traits are measured (see Bulut and Suh (2017) for more details on multidimensional DIF analysis).

11.4   Summary

In the first half of this chapter, we introduced measurement invariance for detecting scale-level bias. Measurement invariance is important to ensure that the construct being measured functions the same way across subgroups of individuals. Violation of invariance can jeopardize further statistical analyses. We described the steps involved in testing measurement invariance (configural, weak, strong, strict, and partial measurement invariance) and demonstrated how to test each step using the **lavaan** package (Rosseel, 2012). In the second half of this chapter, we introduced DIF. We explained two types of DIF (uniform and nonuniform) and described the Mantel-Haenszel (MH) method, logistic regression, and the item response theory likelihood ratio (IRT-LR) test. We demonstrated the MH method and logistic regression using the **difR** package (Magis et al., 2010) and the IRT-LR test using the **mirt** package (Chalmers, 2012). Among the three DIF detection methods introduced here, the IRT-LR test is the most flexible method with regard to item type (i.e., dichotomous versus polytomous) and DIF type (i.e., uniform versus nonuniform). However, it should be noted that the IRT-LR test requires using a pre-specified IRT model (see Chapters 5 through 7 for more details on the IRT models). While not covered in this chapter, the **semTools** (semTools Contributors, 2016) can automate all the measurement invariance steps via its `measurementinvariance` function and the **lordif** package (Choi, Gibbons, & Crane, 2016) offers additional functions for examining DIF for both dichotomous and polytomous items.

12

More Advanced Topics in Measurement

12.1 Chapter Overview

In this concluding chapter, we provide a brief overview of other R packages that may be of interest to researchers and practitioners in education and psychology. These packages focus on advanced psychometric methods that we did not include in this handbook.

12.2 CRAN Task Views

At their core, CRAN task views[1] are meta-packages that allow for a quick and easy installation of a suite of related R packages. Task views are curated and maintained by members of the R community. One installs a task view, which in turn, installs all the packages contained within that particular task view. Therefore, task views allow users to efficiently install and update related packages without needing to individually install packages, though it does not completely remove the need to maintain a separate R script for re-installing desired R packages when updating R or moving to a new computer. In addition, task views are a great way to learn about related packages in the R ecosystem, especially if one's needs involve more esoteric methods.

Task views exist for a multitude of different areas within statistics, including graphics (as mentioned in Chapter 9), Bayesian inference, machine learning, and so on. Of particular interest to readers of this handbook are the CRAN task views for psychometrics (called Psychometrics) and social sciences (called SocialSciences). These task views install many of the packages discussed throughout this book (including **mirt**, **lavaan**, **psych**, and **equate**) as well as packages in various related subfields. For example, the psychometrics task view installs packages related to item response theory, correspondence

---

[1]https://cran.r-project.org/web/views/

analysis and optimal scaling, data reduction (e.g., factor analysis and principal components analysis), structural equation modeling, classical test theory, multidimensional scaling, and more.

To install a task view, we first need to install and load the **ctv** package (Zeileis, 2005).

```
install.packages("ctv")
library("ctv")
```

There are two functions within the **ctv** package that a user is likely to need. To install a task view, such as the Psychometrics task view, we would type

```
install.views("Psychometrics")
```

and to update the packages within this task view, we would type

```
update.views("Psychometrics")
```

There are two important caveats when working with task views. First, sometimes packages are only available for one platform and not another (e.g., a package may only be available for Windows and not for Mac or Linux). Second, installing task views can install hundreds of packages after we factor in the core packages of the task view and their dependencies. So, if disk space or internet bandwidth are an issue, it may be better to install packages individually and maintain a separate R script for package installations.

12.3   Computerized Adaptive Testing

Computerized adaptive testing (CAT) is an advanced test administration system where each examinee receives a unique set of items selected from a large item bank based on the examinee's response pattern (e.g., right or wrong if the items are dichotomously scored). The basic principle behind CAT is that examinees should receive the most appropriate items instead of receiving many items that are too easy or too difficult based on their latent trait levels. This practice results in a test administration that yields highly accurate estimates of the latent trait, with much fewer items compared to paper-and-pencil and computer-based tests that typically consist of many test items regardless of examinees' latent trait levels.

A typical CAT system consists of three steps. First, an examinee begins the test with an item appropriate for a pre-determined latent trait level (e.g., $\theta = 0$), which is also considered the examinee's initial latent trait level. Second, depending on the examinee's response to the item, the CAT system gives an updated estimate of the examinee's latent trait level. Third, based on the updated latent trait level, the CAT system selects the most appropriate item from the item bank. This iterative process continues until a pre-determined test termination criterion is met (e.g., a maximum number of test items, an acceptable level of standard error for the latent trait).

Researchers and practitioners often conduct simulation studies to answer various research questions regarding feasibility, applicability, and planning of CAT. Thompson and Weiss (2011) describe three types of simulation studies in CAT:

1. post-hoc simulations to investigate if CAT would be feasible to use for a paper-and-pencil or computer-based test that has already been administered to a group of examinees,

2. Monte Carlo simulations to evaluate the performances of various CAT approaches under hypothetical testing scenarios, and

3. hybrid simulations to investigate the feasibility of CAT for a paper-and-pencil or computer-based test where each examinee answered at least some of the items in the item bank.

There are several packages available in R to conduct CAT-related simulation studies. For researchers and practitioners, we want to mention two packages, **mirtCAT** (Chalmers, 2016) and **catR** (Magis & Barrada, 2017). Both packages provide researchers and practitioners with a set of simulation functions that are capable of simulating various CAT scenarios with advanced item selection and latent trait estimation algorithms.

The **catR** package consists of routines for generating response patterns for CAT under the unidimensional item response theory (IRT) framework. Both dichotomous and polytomous IRT models can be used in the simulations. The package offers many advanced algorithms for latent trait estimation, optimal item selection, test termination, controlling item exposure, and content balancing. The `randomCAT` is the primary function to design and run simulation studies in the **catR** package. An important feature of the **catR** package is that it allows users to design and implement a real CAT administration using Concerto (http://concertoplatform.com/), which is an open-source platform for administering online and offline adaptive tests.

The **mirtCAT** package allows users to design CAT simulations under the unidimensional and multidimensional IRT frameworks. Like the **catR** package, both dichotomous and polytomous IRT models can be used in the simulations. The package allows users to design both adaptive and non-adaptive tests under various testing scenarios. The `mirtCAT` function is used to design and run simulation studies in the **mirtCAT** package. The `mirtCAT` package also offers additional features - such as generating an HTML interface (using the **shiny** package) for administering live adaptive or non-adaptive tests, running post-hoc and Monte Carlo simulation studies, and visualizing the results from a CAT application.

12.4   Cognitive Diagnostic Modeling

Cognitive diagnostic modeling (CDM), which is also known as diagnostic classification modeling, has been one of the most popular psychometric areas over the past several years. CDM defines examinees' multiple skills or attributes on a discrete scale (e.g., mastery versus non-mastery) as opposed to a continuous scale used for defining a latent trait in the IRT framework. The CDM approaches aim to provide diagnostic information on examinees' strengths and weaknesses based on their mastery status for each of the skills or attributes required for answering the items. For a detailed overview and comparison of various cognitive diagnostic models, we encourage our readers to look over Rupp, Templin, and Henson (2010) and de la Torre and Minchen (2014).

Recently, the CDM framework has been applied to other psychometric research areas, such as adaptive testing. Cognitive diagnostic computer adaptive testing (CD-CAT) combines the efficiency and accuracy of CAT with individualized diagnostic feedback derived from CDM (Cheng, 2009). These advancements have also increased the interest in CDM and its applications in R. Although the number of the CDM packages in R is still growing, we want to mention two packages: **CDM** (George, Robitzsch, Kiefer, Groß, & Ünlü, 2016) and **GDINA** (Ma & de la Torre, 2017). Both packages provide a variety of tools for estimating cognitive diagnostic models for diagnostic assessments with either dichotomous or polytomous items. The **GDINA** package also provides a graphical user interface that allows users to upload their input files and select the desired estimation options for various cognitive diagnostic models (see the `startGDINA` function in **GDINA** for more details).

12.5   IRT Linking Procedures

In Chapter 10, we explained various equating designs (e.g., single group, equivalent group, and non-equivalent group) and equating methods (e.g., general linear and equipercentile) in the context of classical test theory (CTT). The equating process in CTT aims to place test scores from different test administrations or examinee groups onto a common scale so that a meaningful score comparison can be done among the examinees. Unlike in CTT, the equating applications in IRT achieve a common scale for test scores by linking item parameters obtained from different test administrations or examinee groups. Therefore, the equating process in IRT is often called linking. Some IRT linking methods include concurrent calibration, the Stocking-Lord method (Stocking & Lord, 1983), the Haebara method (Haebara, 1980), and the mean/mean and mean/sigma methods (A. S. Cohen & Kim, 1998). For a detailed review of the IRT linking methods, our readers are referred to Lee and Ban (2009) and Kolen and Brennan (2014).

There are several R packages that provide tools for implementing IRT linking procedures, such as **plink** (Weeks, 2010), **equateIRT** (Battauz, 2015), and **SNSequate** (Gonzalez Burgos, 2017). Among these packages, the **plink** package is the most comprehensive one because it enables the linking of mixed-format tests consisting of dichotomous and polytomous items for multiple examinee groups using unidimensional and multidimensional IRT methods. The package supports the mean/sigma, mean/mean, Haebara, and Stocking-Lord linking methods. The `plink` function is the primary function for linking mixed-format tests using unidimensional and multidimensional IRT linking procedures. The package also includes additional functions for the IRT true-score and observed-score equating methods. The `equate` is the primary function for conducting IRT true score and observed score equating for mixed-format tests under the unidimensional IRT framework. We encourage our readers to look over the **plink** article in the *Journal of Statistical Software* (Weeks, 2010) for more examples of linking with **plink**.

12.6 Bayesian Models of Measurement

Traditional statistical methods employ a frequentist perspective of probability. In other words, the probability of an event occurring is the relative frequency of that event happening over an infinitely large number of replicated studies, experiments, test administrations, etc. In contrast to the frequentist perspective, the Bayesian perspective defines probability in terms of degree of belief and represents a subjective view of probability.

Conceptually, Bayesian statistics derives a probability distribution, known as the posterior distribution, that is approximately the product of this subjective view (known as the prior distribution) and the likelihood function (the statistical model typically fit in frequentist statistics) via Bayes theorem. The issue of what prior distribution to specify, if any, has been hotly debated as well as whether a prior distribution can ever be non-informative. While all models are subjective, Bayesian modeling has been ridiculed in the past as being too subjective because of these reasons. Fortunately, this debate has largely subsided and the focus is now on the utility of Bayesian methods.

Because many posterior distributions cannot be derived analytically but instead must be approximated using Markov Chain Monte Carlo methods (Gelman et al., 2014), which are quite computationally taxing, these methods have only recently started gaining popularity in applied fields as computing power has increased and the desire to solve more complex statistical problems has grown. Many statistical models simply cannot be easily fit within the traditional modeling framework but can be readily fit within a Bayesian model because of the use of the prior distribution.

While we have presented a frequentist perspective throughout this handbook, we do not intend to indicate a bias or preference over frequentist models. Instead, we believe that frequentist models should be thoroughly understood before Bayesian models are considered. The reason is that many Bayesian models assume default prior distributions, which may or may not be reasonable for our applications, and it is important to understand our model before adding this extra level of complexity; otherwise we would add an extra layer to an already large "black box." Also, while it would be debatable whether an understanding of statistical theory is necessary to fit frequentist models in an applied setting, given that Bayesian models require specification of probability distributions, at least a rudimentary knowledge of statistical theory is necessary. For readers with strong mathematical backgrounds, we recommend either Gelman et al. (2014) or Carlin and Louis (2008). For readers requiring a more gentle introduction with minimal statistical theory, Kruschke (2014) would be a good start.

Within R, there are a plethora of engines and libraries to fit Bayesian measurement models. For advanced Bayesian users, the **rjags** (Plummer, 2016), **runjags** (Denwood, 2016), and **rstan** (Stan Development Team, 2016) packages provide R interfaces to the JAGS and Stan modeling platforms. These libraries act as wrapper functions and allow users to more easily interact with these Bayesian programming languages. The drawback of using these platforms is that users would need to know how to write the syntax for their models. The advantage, however, is that once users are able to write their model, they can easily fit any measurement model within a Bayesian framework using these libraries.

Many readers of this handbook may find these libraries intimidating. Fortunately, there are other more user-friendly Bayesian packages. To fit confirmatory Bayesian latent variable models, the **blavaan** package (Merkle & Rosseel, 2016) can be used, while Bayesian exploratory factor models can be fit using either the **BayesFM** (Piatek, 2017) or **MCMCpack** (Martin, Quinn, & Park, 2011) packages. In addition, the **MCMCpack** and **pscl** (Jackman, 2015) packages can be used to fit IRT models. Finally, the **edstan** package

(Furr, 2017) is a convenient package for estimating some IRT models (e.g., Rasch, 2PL, partial credit, and rating scale models) for the Stan language, and can be used to interface with Stan with some limitations.

Many of these packages differ in their flexibility of specifying the prior distributions. While these specifications may be appropriate for most analyses, we strongly recommend sensitivity analyses, where users can vary the prior distributions, as a way to understand and quantify the magnitude of the impact of the prior distribution(s) on the parameter estimates and model findings.

12.7 Hierarchical Linear Models

While not a measurement model, hierarchical linear models[2] (HLMs) are common and extremely important statistical models in education and the social sciences. When data are collected repeatedly on participants or when data are collected on students nested within classrooms that are nested within schools, schools districts and so on, these observations will not be independent, and the residuals within a participant or across students within a classroom will be correlated. HLMs allow us to statistically control for this lack of independence, which would otherwise affect our standard errors and estimated parameters. In addition, while we might administer an assessment to a random sample of teachers, we might not be interested solely in those teachers (i.e., treating teachers as a fixed effect) but instead all teachers (i.e., treating teachers as a random effect). This can be done within an HLM.[3]

The generalizability theory models presented in Chapter 3 and the explanatory IRT models presented in Chapter 8 are really just HLMs in disguise. This is why a single, commonly utilized R package for HLM, **lme4** (Bates et al., 2015), was used in both instances. The **lme4** package is capable of fitting both linear mixed effects (when the outcome variable is assumed to be conditionally normally distributed) and generalized linear mixed effects (when the outcome variable could have a binomial, Poisson, gamma, or other distribution, see `?family` for a list of potential distributions). The main functions in **lme4** are `lmer`, for fitting a linear mixed effects model, and `glmer`, for fitting a generalized linear mixed effects model. Both functions require at least a single random effect, otherwise either general linear (e.g., multiple regression) or generalized linear regression (e.g., logistic regression) need to be performed. In addition to **lme4**, the **nlme** package (Pinheiro, Bates, DebRoy, Sarkar, &

---

[2]This is also known as mixed effects, random effects, and multilevel modeling.
[3]This may also be done in a random effects ANOVA.

R Core Team, 2017), typically included with a base R installation, can fit linear mixed effects models as well as more complex error structures that **lme4** cannot.

Practitioners using **lme4** might be dismayed to realize that p-values are not available for their models, while they are available from **nlme**. However, there are several options for calculating them in HLMs.[4] Also, the **lmerTest** package (Kuznetsova, Bruun Brockhoff, & Haubo Bojesen Christensen, 2016) is capable of producing p-values for the parameter estimates obtained via the **lme4** package (particularly with the `lmer` and `glmer` functions). Regardless, this should not weigh into the decision to choose either **nlme** or **lme4**. We typically use **lme4** because it is actively being developed, and only move to **nlme** when we need to fit a model that we cannot estimate with **lme4**.

Finally, there are many excellent books on fitting these models in R. This includes the mixed effects book written by the authors of **nlme** and one of the authors of **lme4** (Pinheiro & Bates, 2006). While the book was written for the **nlme** library in S, the precursor of R, the code and statistical methods described in the text are the same for R and are still applicable today. The general regression modeling books by Faraway (2014, 2016) are another great resource for HLMs. In Faraway (2016), the author describes HLMs as well as generalized linear modeling and nonparametric regression. Finally, the all-purpose regression and data analysis book written by Gelman and Hill (2007) is an excellent resource for practitioners that need to do data analysis; that need to know how to and why to run simulations; and that want to run HLMs in R. Gelman and Hill (2007) move quickly, starting with the basics and moving through Bayesian HLM using R, but it is highly readable and packed with lots of practical advice for researchers and practitioners.

12.8   Profile Analysis

Profile analysis is a technique that is widely used by researchers in education, psychology, and medicine for the non-orthogonal decomposition of observed scores into level and pattern effects. The **profileR** package (Bulut & Desjardins, 2017) in R includes routines to perform criterion-related profile analysis (M. L. Davison & Davenport, 2002), profile analysis via multidimensional scaling (M. L. Davison, 1994), moderated profile analysis, profile analysis by group, profile reliability (Bulut, 2013; Bulut et al., 2017), and within-person score profiles derived using the factor model (M. L. Davison, Kim, & Close, 2009). The **profileR** package allows researchers to quantify variability associ-

---

[4]https://bbolker.github.io/mixedmodels-misc/glmmFAQ.html

ated with a particular effect (i.e., level or pattern), test a series of hypotheses about a profile, quickly create profile plots, and extract information for further analyses (e.g., regression).

The primary functions in the **profileR** package include `cpa` for criterion-related profile analysis, `mpa` for moderated profile analysis, `pams` for profile analysis via multidimensional scaling, `pbg` for profile analysis by group, `pr` for within-person and between-person profile reliability, and `profileplot` for drawing a profile plot. We encourage our readers to check out the package manual for further details on the **profileR** package.

12.9 Summary

In this chapter, we presented an overview R packages of interest to researchers and practitioners in education and psychology not discussed in this book. We discussed CRAN task views, the **catR** and **mirtCAT** packages for conducting simulation studies in CAT, the **CDM** and **GDINA** packages for estimating cognitive diagnostic models, the **plink** package for linking procedures in uni-dimensional and multidimensional IRT models, the **lme4** and **nlme** packages for estimating hierarchical linear and nonlinear models, and the **profileR** package for conducting profile analysis. In addition, we mentioned several R packages that are capable of estimating Bayesian models and interacting wtih other Bayesian software programs such as JAGS and Stan.

References

Ackerman, T. A., Gierl, M. J., & Walker, C. M. (2003). Using multidimensional item response theory to evaluate educational and psychological tests. *Educational Measurement: Issues and Practice*, *22*(3), 37–51.

Adams, R. J., Wilson, M., & Wang, W.-C. (1997). The multidimensional random coefficients multinomial logit model. *Applied Psychological Measurement*, *21*(1), 1–23. doi: 10.1177/0146621697211001

Agresti, A. (2002). *Categorical data analysis* (2nd ed.). Hoboken, NJ: John Wiley & Sons.

Albano, A. D. (2015). A general linear method for equating with small samples. *Journal of Educational Measurement*, *52*(1), 55–69.

Albano, A. D. (2016). equate: An R package for observed-score linking and equating. *Journal of Statistical Software*, *74*(8), 1–36. doi: 10.18637/jss.v074.i08

Andrich, D. (1978). A rating formulation for ordered response categories. *Psychometrika*, *43*(4), 561–573. doi: 10.1007/BF02293814

Asparouhov, T., & Muthén, B. (2010). *Simple second order chi-square correction*. Retrieved from https://www.statmodel.com/download/WLSMV_new_chi21.pdf

Bache, S. M., & Wickham, H. (2014). magrittr: A forward-pipe operator for r [Computer software manual]. Retrieved from https://CRAN.R-project.org/package=magrittr (R package version 1.5)

Baker, F. B., & Kim, S.-H. (2017). *The basics of item response theory using R*. Springer.

Bartholomew, D. J., Deary, I. J., & Lawn, M. (2009). The origin of factor scores: Spearman, thomson and bartlett. *British Journal of Mathematical and Statistical Psychology*, *62*(3), 569–582. doi: 10.1348/000711008X365676

Barton, M. A., & Lord, F. M. (1981). An upper asymptote for the three-parameter logistic item-response model. *ETS Research Report Series*(1), 1–8. doi: 10.1002/j.2333-8504.1981.tb01255.x

Bates, D., Mächler, M., Bolker, B., & Walker, S. (2015). Fitting linear mixed-effects models using lme4. *Journal of Statistical Software*, *67*(1), 1–48. doi: 10.18637/jss.v067.i01

Battauz, M. (2015). equateIRT: An R package for irt test equating. *Journal of Statistical Software*, *68*(7), 1–22. doi: 10.18637/jss.v068.i07

Beaujean, A. A. (2014). *Latent variable modeling using R: A step-by-step guide*. Routledge.

Benjamini, Y., & Hochberg, Y. (1995). Controlling the false discovery rate: A practical and powerful approach to multiple testing. *Journal of the Royal Statistical Society. Series B (Methodological)*, *57*(1), 289–300.

Bernaards, C. A., & Jennrich, R. I. (2005). Gradient projection algorithms and software for arbitrary rotation criteria in factor analysis. *Educational and Psychological Measurement*, *65*, 676–696.

Bock, D. R. (1972). Estimating item parameters and latent ability when responses are scored in two or more nominal categories. *Psychometrika*, *37*(1), 29–51. doi: 10.1007/BF02291411

Bollen, K. A. (1989). *Structural equations with latent variables*. John Wiley & Sons.

Bollen, K. A., & Arminger, G. (1991). Observational residuals in factor analysis and structural equation models. *Sociological Methodology*, *21*, 235–262. doi: 10.2307/270937

Braun, H. I., & Holland, P. W. (1982). Observed-score test equating: A mathematical analysis of some ETS equating procedures. In P. W. Holland & D. B. Rubin (Eds.), *Test equating* (pp. 9–49). New York, NY: Academic Press.

Brennan, R. L. (1992, December). Generalizability theory. *Educational Measurement: Issues and Practice*, *11*(4), 27–34. doi: 10.1111/j.1745-3992 .1992.tb00260.x

Brennan, R. L. (2006). *Educational measurement* (4th ed.). Westport, CT: Praeger Publishers.

Briggs, D. C. (2008). Using explanatory item response models to analyze group differences in science achievement. *Applied Measurement in Education*, *21*(2), 89–118.

Bulut, O. (2013). *Between-person and within-person subscore reliability: Comparison of unidimensional and multidimensional IRT models*. (Unpublished doctoral dissertation). University of Minnesota.

Bulut, O., Davison, M. L., & Rodriguez, M. C. (2017). Estimating between-person and within-person subscore reliability with profile analysis. *Multivariate Behavioral Research*, *52*(1), 86–104. doi: 10.1080/00273171 .2016.1253452

Bulut, O., & Desjardins, C. D. (2017). profileR: Profile analysis of multivariate data in R [Computer software manual]. Retrieved from `https://CRAN .R-project.org/package=profileR` (R package version 0.3-4)

Bulut, O., & Suh, Y. (2017). Detecting multidimensional differential item functioning with the multiple indicators multiple causes model, the item response theory likelihood ratio test, and logistic regression. *Frontiers in Education*, *2*(51), 1–14. doi: 10.3389/feduc.2017.00051

Canty, A. (2002). Resampling methods in R: The boot package. *R News*, *2/3*, 2–7.

Canty, A., & Ripley, B. D. (2017). boot: Bootstrap R (S-plus) functions [Computer software manual]. (R package version 1.3-19)

Carlin, B. P., & Louis, T. A. (2008). *Bayesian methods for data analysis* (3rd ed.). Boca Raton, FL: CRC Press.

Chalmers, R. P. (2012). mirt: A multidimensional item response theory package for the R environment. *Journal of Statistical Software, 48*(6), 1–29.

Chalmers, R. P. (2016). Generating adaptive and non-adaptive test interfaces for multidimensional item response theory applications. *Journal of Statistical Software, 71*(5), 1–39. doi: 10.18637/jss.v071.i05

Chalmers, R. P., & Flora, D. B. (2015). faoutlier: An R package for detecting influential cases in exploratory and confirmatory factor analysis. *Applied Psychological Measurement, 39*(7), 573–574.

Chang, W., Cheng, J., Allaire, J., Xie, Y., & McPherson, J. (2017). shiny: Web application framework for R [Computer software manual]. Retrieved from https://CRAN.R-project.org/package=shiny (R package version 1.0.3)

Cheng, Y. (2009). When cognitive diagnosis meets computerized adaptive testing: CD-CAT. *Psychometrika, 74*(4), 619–632. doi: 10.1007/s11336-009-9123-2

Choi, S. W., Gibbons, L. E., & Crane, P. K. (2016). lordif: Logistic ordinal regression differential item functioning using irt [Computer software manual]. Retrieved from https://CRAN.R-project.org/package=lordif (R package version 0.3-3)

Cleveland, W. S. (1985). *The elements of graphing data.* Wadsworth Advanced Books and Software Monterey, CA.

Cleveland, W. S. (1993). *Visualizing data.* Hobart Press.

Cohen, A. S., & Kim, S.-H. (1998). An investigation of linking methods under the graded response model. *Applied Psychological Measurement, 22*(2), 116–130. doi: 10.1177/01466216980222002

Cohen, R. J., Swerdlik, M. E., & Sturman, E. D. (2013). *Psychological testing and assessment: An introduction to tests and measurement* (8th ed.). New York, NY: McGraw-Hall.

Cook, R. D. (1977). Detection of influential observation in linear regression. *Technometrics, 19*(1), 15–18.

Crawley, M. J. (2013). *The R book* (2nd ed.). John Wiley & Sons.

Cronbach, L. J. (1951). Coefficient alpha and the internal structure of tests. *Psychometrika, 16*(3), 297–334. doi: 10.1007/BF02310555

Davison, A. C., & Hinkley, D. V. (1997). *Bootstrap methods and their applications.* Cambridge: Cambridge University Press. (ISBN 0-521-57391-2)

Davison, M. L. (1994). Multidimensional scaling models of personality responding. In S. Strack & M. Lorr (Eds.), *Differentiating normal and abnormal personality.* Springer Publishing Co.

Davison, M. L., & Davenport, E. C. (2002). Identifying criterion-related patterns of predictor scores using multiple regression. *Psychological Methods, 7*(4), 468–484.

Davison, M. L., Kim, S. K., & Close, C. (2009). Factor analytic modeling of within person variation in score profiles. *Multivariate Behavioral*

Research, 44, 668–687.

De Ayala, R. J. (2013). *The theory and practice of item response theory.* New York, NY: The Guilford Press.

De Boeck, P. (2008). Random item IRT models. *Psychometrika, 73*(4), 533–559.

De Boeck, P., Bakker, M., Zwitser, R., Nivard, M., Hofman, A., Tuerlinckx, F., & Partchev, I. (2011). The estimation of item response models with the lmer function from the lme4 package in R. *Journal of Statistical Software, 39*(12), 1–28.

De Boeck, P., & Wilson, M. (2004). *Explanatory item response models.* New York: Springer-Verlag.

de la Torre, J., & Minchen, N. (2014). Cognitively diagnostic assessments and the cognitive diagnosis model framework. *Psicologia Educativa, 20*(2), 89–97. doi: https://doi.org/10.1016/j.pse.2014.11.001

de la Torre, J., Song, H., & Hong, Y. (2011). A comparison of four methods of IRT subscoring. *Applied Psychological Measurement, 35*(4), 296–316. doi: 10.1177/0146621610378653

Denwood, M. J. (2016). runjags: An R package providing interface utilities, model templates, parallel computing methods and additional distributions for MCMC models in JAGS. *Journal of Statistical Software, 71*(9), 1–25. doi: 10.18637/jss.v071.i09

Doran, H., Bates, D., Bliese, P., & Dowling, M. (2007). Estimating the multilevel Rasch model: With the lme4 package. *Journal of Statistical Software, 20*(2), 1–18.

Drasgow, F., Levine, M. V., & Williams, E. A. (1985). Appropriateness measurement with polychotomous item response models and standardized indices. *British Journal of Mathematical and Statistical Psychology, 38*(1), 67–86. doi: 10.1111/j.2044-8317.1985.tb00817.x

Efron, B., & Tibshirani, R. (1986). Bootstrap methods for standard errors, confidence intervals, and other measures of statistical accuracy. *Statistical Science, 1*(1), 54–77.

Embretson, S. E. (1997). Multicomponent response models. In W. J. van der Linden & R. K. Hambleton (Eds.), *Handbook of modern item response theory* (pp. 305–321). New York, NY: Springer.

Embretson, S. E., & Reise, S. P. (2000). *Item response theory for psychologists.* Mahwah, NJ: Lawrence Erlbaum Associates.

Epskamp, S., & Stuber, S. (2017). semPlot: Path diagrams and visual analysis of various SEM packages' output [Computer software manual]. Retrieved from https://CRAN.R-project.org/package=semPlot (R package version 1.1)

Falissard, B. (2012). *Analysis of questionnaire data with R.* Boca Raton, FL: CRC Press.

Faraway, J. J. (2014). *Linear models with R.* Boca Raton, FL: CRC Press.

Faraway, J. J. (2016). *Extending the linear model with R: Generalized linear, mixed effects and nonparametric regression models* (Vol. 124). Boca

Raton, FL: CRC Press.

Finch, W. H., & French, B. F. (2015). *Latent variable modeling with R.* Routledge.

Fischer, G. H. (1973). The linear logistic test model as an instrument in educational research. *Acta Psychologica, 37*(6), 359–374.

Flora, D. B., LaBrish, C., & Chalmers, R. P. (2012). Old and new ideas for data screening and assumption testing for exploratory and confirmatory factor analysis. *Frontiers in Psychology, 3.* doi: 10.3389/fpsyg.2012 .00055

Fox, J., Nie, Z., & Byrnes, J. (2017). sem: Structural equation models [Computer software manual]. Retrieved from https://CRAN.R-project .org/package=sem (R package version 3.1-9)

Furr, D. C. (2017). edstan: Stan models for item response theory [Computer software manual]. Retrieved from https://CRAN.R-project .org/package=edstan (R package version 1.0.6)

Gelman, A., Carlin, J. B., Stern, H. S., Dunson, D. B., Vehtari, A., & Rubin, D. B. (2014). *Bayesian data analysis* (3rd ed., Vol. 2). Boca Raton, FL: CRC Press.

Gelman, A., & Hill, J. (2007). *Data analysis using regression and multilevel hierarchical models* (Vol. 1). Cambridge University Press New York, NY, USA.

George, A. C., Robitzsch, A., Kiefer, T., Groß, J., & Ünlü, A. (2016). The R package CDM for cognitive diagnosis models. *Journal of Statistical Software, 74*(2), 1–24. doi: 10.18637/jss.v074.i02

Gierl, M. J., Bulut, O., Guo, Q., & Zhang, X. (2017). Developing, analyzing, and using distractors for multiple-choice tests in education: A comprehensive review. *Review of Educational Research, 87*(6), 1082–1116. doi: 10.3102/0034654317726529

Gonzalez Burgos, J. (2017). SNSequate: Standard and nonstandard statistical models and methods for test equating [Computer software manual]. Retrieved from https://CRAN.R-project.org/package=SNSequate (R package version 1.3.1)

Gorsuch, R. L. (1983). *Factor analysis* (2nd ed.). Hillsdale, NJ: Lawrence Erlbaum Associates, Inc.

Haebara, T. (1980). Equating logistic ability scales by a weighted least squares method. *Japanese Psychological Research, 22*(3), 144–149. doi: 10.4992/ psycholres1954.22.144

Hambleton, R. K., Swaminathan, H., & Rogers, H. J. (1991). *Fundamentals of item response theory.* Newbury Park, CA: Sage.

Holland, P. W., & Thayer, D. T. (1988). Differential item performance and the mantel-haenszel procedure. In H. Wainer & H. I. Braun (Eds.), *Test validity* (pp. 129–145). Hillsdale, NJ: Lawrence Erlbaum Associates.

Horn, J. L. (1965). A rationale and test for the number of factors in factor analysis. *Psychometrika, 30*(2), 179–185. doi: 10.1007/BF02289447

Hu, L.-t., & Bentler, P. M. (1999). Cutoff criteria for fit indexes in covariance structure analysis: Conventional criteria versus new alternatives. *Structural equation modeling: A multidisciplinary journal*, *6*(1), 1–55.

Jackman, S. (2015). pscl: Classes and methods for R developed in the political science computational laboratory, stanford university [Computer software manual]. Stanford, California. Retrieved from `http://pscl.stanford.edu/` (R package version 1.4.9)

Jeon, M., Rijmen, F., & Rabe-Hesketh, S. (2014). Flexible item response theory modeling with flirt. *Applied Psychological Measurement*, *38*(5), 404–405.

Jodoin, M. G., & Gierl, M. J. (2001). Evaluating type I error and power rates using an effect size measure with the logistic regression procedure for DIF detection. *Applied Measurement in Education*, *14*(4), 329–349. doi: 10.1207/S15324818AME1404_2

Jöreskog, K. G., Sörbom, D., & Wallentin, F. Y. (2006). Latent variable scores and observational residuals. *Retrieved June*, *7*, 2009.

Junker, B. W., & Sijtsma, K. (2001). Cognitive assessment models with few assumptions, and connections with nonparametric item response theory. *Applied Psychological Measurement*, *25*(3), 258–272. doi: 10.1177/01466210122032064

Kaiser, H. F. (1960). The application of electronic computers to factor analysis. *Educational and Psychological Measurement*, *20*(1), 141–151. doi: 10.1177/001316446002000116

Kiefer, T., Robitzsch, A., & Wu, M. (2016). TAM: Test analysis modules [Computer software manual]. Retrieved from `https://CRAN.R-project.org/package=TAM` (R package version 1.995-0)

Kline, R. B. (2015). *Principles and practice of structural equation modeling.* Guilford Publications.

Kolen, M. J., & Brennan, R. L. (2014). *Test equating, scaling, and linking.* New York, NY: Springer.

Korkmaz, S., Goksuluk, D., & Zararsiz, G. (2014). MVN: An R package for assessing multivariate normality. *The R Journal*, *6*(2), 151–162. Retrieved from `http://journal.r-project.org/archive/2014-2/korkmaz-goksuluk-zararsiz.pdf`

Kruschke, J. (2014). *Doing Bayesian data analysis: A tutorial with R, JAGS, and Stan.* Academic Press.

Kuznetsova, A., Bruun Brockhoff, P., & Haubo Bojesen Christensen, R. (2016). lmerTest: Tests in linear mixed effects models [Computer software manual]. Retrieved from `https://CRAN.R-project.org/package=lmerTest` (R package version 2.0-33)

Lane, S., Raymond, M. R., & Haladyna, T. M. (2015). *Handbook of test development* (2nd ed.). New York, NY: Routledge.

Lawshe, C. H. (1975). A quantitative approach to content validity. *Personnel Psychology*, *28*(4), 563–575.

Lê, S., Josse, J., & Husson, F. (2008). FactoMineR: A package for multivariate analysis. *Journal of Statistical Software*, *25*(1), 1–18. doi: 10.18637/

jss.v025.i01

Lee, W.-C., & Ban, J.-C. (2009). A comparison of IRT linking procedures. *Applied Measurement in Education*, *23*(1), 23–48. doi: 10.1080/08957340903423537

Linacre, J. M. (2015). *A user's guide to WINSTEPS MINISTEP Rasch-model computer programs*. Chicago, IL: Winsteps.com.

Livingston, S. A. (2014). Equating test scores (without IRT). *Educational Testing Service*. Retrieved 2017-05-15, from http://eric.ed.gov/?id=ED560972

Loken, E., & Rulison, K. L. (2010). Estimation of a four-parameter item response theory model. *British Journal of Mathematical and Statistical Psychology*, *63*(3), 509–525. doi: 10.1348/000711009X474502

Lord, F. M. (1980). *Applications of item response theory to practical testing problems*. Hillsdale, NJ: Lawrence Erlbaum Associates.

Ma, W., & de la Torre, J. (2017). GDINA: The generalized DINA model framework [Computer software manual]. Retrieved from https://CRAN.R-project.org/package=GDINA (R package version 1.4.2)

Magis, D. (2013). A note on the item information function of the four-parameter logistic model. *Applied Psychological Measurement*, *37*(4), 304-315. doi: 10.1177/0146621613475471

Magis, D., & Barrada, J. R. (2017). Computerized adaptive testing with R: Recent updates of the package catR. *Journal of Statistical Software, Code Snippets*, *76*(1), 1–19. doi: 10.18637/jss.v076.c01

Magis, D., Beland, S., Tuerlinckx, F., & De Boeck, P. (2010). A general framework and an R package for the detection of dichotomous differential item functioning. *Behavior Research Methods*, *42*, 847–862.

Mair, P., & Hatzinger, R. (2007a). eRm: Extended rasch modeling: The eRm package for the application of IRT models in R. *Journal of Statistical Software*, *20*(9), 1–20. Retrieved from http://www.jstatsoft.org/v20/i09/

Mair, P., & Hatzinger, R. (2007b). Psychometrics task view. *R News*, *7*(3), 38–40. Retrieved from https://cran.r-project.org/doc/Rnews/Rnews_2007-3.pdf

Mantel, N., & Haenszel, W. (1959). Statistical aspects of the analysis of data from retrospective studies of disease. *Journal of the National Cancer Institute*, *22*(4), 719–748.

Maris, E. (1995). Psychometric latent response models. *Psychometrika*, *60*(4), 523–547. doi: 10.1007/BF02294327

Martin, A. D., Quinn, K. M., & Park, J. H. (2011). MCMCpack: Markov Chain Monte Carlo in R. *Journal of Statistical Software*, *42*(9), 22. Retrieved from http://www.jstatsoft.org/v42/i09/

Masters, G. N. (1982). A Rasch model for partial credit scoring. *Psychometrika*, *47*(2), 149–174. doi: 10.1007/BF02296272

Mavridis, D., & Moustaki, I. (2008). Detecting outliers in factor analysis using the forward search algorithm. *Multivariate Behavioral Research*, *43*(3),

453–475.

Maydeu-Olivares, A. (2013). Goodness-of-fit assessment of item response theory models. *Measurement: Interdisciplinary Research and Perspectives*, *11*(3), 71–101. doi: 10.1080/15366367.2013.831680

McKinley, R. L., & Mills, C. N. (1985). A comparison of several goodness-of-fit statistics. *Applied Psychological Measurement*, *9*(1), 49–57. doi: 10.1177/014662168500900105

Merkle, E. C., & Rosseel, Y. (2016). blavaan: Bayesian structural equation models via parameter expansion. *arXiv 1511.05604*. Retrieved from https://arxiv.org/abs/1511.05604

Moore, C. T. (2016). gtheory: Apply generalizability theory with R [Computer software manual]. Retrieved from https://CRAN.R-project.org/package=gtheory (R package version 0.1.2)

Mulaik, S. A. (2009). *Foundations of factor analysis*. Boca Raton, FL: CRC Press.

Muraki, E. (1990). Fitting a polytomous item response model to Likert-type data. *Applied Psychological Measurement*, *14*(1), 59–71.

Muraki, E. (1992). A generalized partial credit model: Application of an EM algorithm. *Applied Psychological Measurement*, *16*(2), 159-176. doi: 10.1177/014662169201600206

Muraki, E., & Bock, R. D. (1997). *PARSCALE: IRT item analysis and test scoring for rating-scale data*. Chicago, IL: Scientific Software International.

Muthén, L., & Muthén, B. (2015). *Mplus user's guide* (7th ed.). Muthén & Muthén.

Neale, M. C., Hunter, M. D., Pritikin, J. N., Zahery, M., Brick, T. R., Kirkpatrick, R. M., ... Boker, S. M. (2016). OpenMx 2.0: Extended structural equation and statistical modeling. *Psychometrika*, *81*(2), 535–549. doi: 10.1007/s11336-014-9435-8

Orlando, M., & Thissen, D. (2000). Likelihood-based item-fit indices for dichotomous item response theory models. *Applied Psychological Measurement*, *24*(1), 50–64. doi: 10.1177/01466216000241003

Osgood, D. W., McMorris, B. J., & Potenza, M. T. (2002). Analyzing multiple-item measures of crime and deviance I: Item response theory scaling. *Journal of Quantitative Criminology*, *18*(3), 267–296. doi: 10.1023/A: 1016008004010

Pek, J., & MacCallum, R. C. (2011). Sensitivity analysis in structural equation models: Cases and their influence. *Multivariate Behavioral Research*, *46*(2), 202–228.

Penfield, R. D., & Camilli, G. (2006). Differential item functioning and item bias. In C. R. Rao & S. Sinharay (Eds.), *Psychometrics* (Vol. 26, pp. 125–167). Elsevier. doi: https://doi.org/10.1016/S0169-7161(06)26005-X

Piatek, R. (2017). BayesFM: Bayesian inference for factor modeling [Computer software manual]. Retrieved from https://CRAN.R-project.org/package=BayesFM (R package version 0.1.2)

Pinheiro, J., & Bates, D. (2006). *Mixed-effects models in S and S-PLUS.* Springer Science & Business Media.

Pinheiro, J., Bates, D., DebRoy, S., Sarkar, D., & R Core Team. (2017). nlme: Linear and nonlinear mixed effects models [Computer software manual]. Retrieved from https://CRAN.R-project.org/package=nlme (R package version 3.1-131)

Plummer, M. (2016). rjags: Bayesian graphical models using MCMC [Computer software manual]. Retrieved from https://CRAN.R-project.org/package=rjags (R package version 4-6)

R Core Team. (2017). R: A language and environment for statistical computing [Computer software manual]. Vienna, Austria. Retrieved from https://www.R-project.org/

Rasch, G. (1960). *Probabilistic models for some intelligence and attainment tests.* Copenhagen: Danish Institute for Educational Research.

Reckase, M. D. (1985). The difficulty of test items that measure more than one ability. *Applied Psychological Measurement, 9*(4), 401–412. doi: 10.1177/014662168500900409

Reckase, M. D. (1997). The past and future of multidimensional item response theory. *Applied Psychological Measurement, 21*(1), 25–36. doi: 10.1177/0146621697211002

Reckase, M. D. (2009). *Multidimensional item response theory* (1st ed.). New York, NY: Springer-Verlag.

Reise, S. P., & Revicki, D. A. (2014). *Handbook of item response theory modeling: Applications to typical performance assessment.* Taylor & Francis.

Reise, S. P., & Waller, N. G. (2009). Item response theory and clinical measurement. *Annual Review of Clinical Psychology, 5*(1), 27–48. doi: 10.1146/annurev.clinpsy.032408.153553

Revelle, W. (2017). psych: Procedures for psychological, psychometric, and personality research [Computer software manual]. Evanston, Illinois. Retrieved from https://CRAN.R-project.org/package=psych (R package version 1.7.3)

Rizopoulos, D. (2006). ltm: An R package for latent variable modelling and item response theory analyses. *Journal of Statistical Software, 17*(5), 1–25. Retrieved from http://www.jstatsoft.org/v17/i05/

Robitzsch, A. (2017). sirt: Supplementary item response theory models [Computer software manual]. Retrieved from https://CRAN.R-project.org/package=sirt (R package version 2.0-25)

Rosenberg, M. (1965). *Society and the adolescent self-image.* Princeton, NJ: Princeton University Press.

Rosseel, Y. (2012). lavaan: An R package for structural equation modeling. *Journal of Statistical Software, 48*(2), 1–36. Retrieved from http://www.jstatsoft.org/v48/i02/

Rupp, A. A., Templin, J., & Henson, R. A. (2010). *Diagnostic measurement: Theory, methods, and applications.* New York, NY: Guilford Press.

Rupp, A. A., & Zumbo, B. D. (2006). Understanding parameter invariance in unidimensional IRT models. *Educational and Psychological Measurement, 66*(1), 63–84. doi: 10.1177/0013164404273942

Samejima, F. (1969). *Estimation of latent ability using a response pattern of graded scores* (Vol. 17). Richmond, VA: Psychometric Society.

Sarkar, D. (2008). *Lattice: Multivariate data visualization with R.* New York: Springer. Retrieved from http://lmdvr.r-forge.r-project.org (ISBN 978-0-387-75968-5)

semTools Contributors. (2016). semTools: Useful tools for structural equation modeling [Computer software manual]. Retrieved from https://CRAN.R-project.org/package=semTools (R package version 0.4-14)

Shavelson, R. J., & Webb, N. M. (1991). *Generalizability theory: A primer.* Sage.

Sijtsma, K., & Junker, B. W. (2006). Item response theory: Past performance, present developments, and future expectations. *Behaviormetrika, 33*(1), 75–102. doi: 10.2333/bhmk.33.75

Stan Development Team. (2016). *RStan: The R interface to Stan.* Retrieved from http://mc-stan.org/ (R package version 2.14.1)

Stocking, M. L., & Lord, F. M. (1983). Developing a common metric in item response theory. *Applied Psychological Measurement, 7*(2), 201–210. doi: 10.1177/014662168300700208

Strout, W. F. (1990). A new item response theory modeling approach with applications to unidimensionality assessment and ability estimation. *Psychometrika, 55*(2), 293–325. doi: 10.1007/BF02295289

Suh, Y., & Bolt, D. M. (2010). Nested logit models for multiple-choice item response data. *Psychometrika, 75*(3), 454–473.

Sveidqvist, K., Bostock, M., Pettitt, C., Daines, M., Kashcha, A., & Iannone, R. (2017). DiagrammeR: Create graph diagrams and flowcharts using R [Computer software manual]. Retrieved from https://CRAN.R-project.org/package=DiagrammeR (R package version 0.9.0)

Swaminathan, H., Hambleton, R. K., & Rogers, H. J. (2006). 21 assessing the fit of item response theory models. In C. R. Rao & S. Sinharay (Eds.), *Psychometrics* (Vol. 26, pp. 683–718). Elsevier. doi: 10.1016/S0169-7161(06)26021-8

Swaminathan, H., & Rogers, H. J. (1990). Detecting differential item functioning using logistic regression procedures. *Journal of Educational Measurement, 27*(4), 361–370.

Tabachnick, B. G., Fidell, L. S., & Osterlind, S. J. (2001). *Using multivariate statistics* (3rd ed.). Boston, MA: Allyn and Bacon.

Tavares, H. A. R., Andrade, D. F. d., & Pereira, C. A. d. B. A. (2004). Detection of determinant genes and diagnostic via item response theory. *Genetics and Molecular Biology, 27*, 679–685.

Thissen, D., Pommerich, M., Billeaud, K., & Williams, V. S. L. (1995). Item response theory for scores on tests including polytomous items with ordered responses. *Applied Psychological Measurement, 19*(1), 39–49. doi: 10.1177/014662169501900105

Thissen, D., Steinberg, L., & Wainer, H. (1993). Detection of differential item functioning using the parameters of item response models. In P. W. Holland & H. Wainer (Eds.), *Differential item functioning* (pp. 67–113). Hillsdale, NJ: Lawrence Erlbaum Associates.

Thompson, N. A., & Weiss, D. J. (2011). A framework for the development of computerized adaptive tests. *Practical Assessment, Research & Evaluation, 16*(1), 1–9.

Thorndike, R. M., & Thorndike-Christ, T. M. (2010). *Measurement and evaluation in psychology and education* (8th ed.). New York: Pearson.

Vansteelandt, K. (2000). *Formal models for contextualized personality psychology* (Unpublished doctoral dissertation). KU Leuven, Belgium.

Venables, W. N., & Ripley, B. D. (2002). *Modern applied statistics with S* (Fourth ed.). New York: Springer. (ISBN 0-387-95457-0)

Venables, W. N., & Smith, D. M. (2016). *An Introduction to R. Notes on R: A Programming Environment for Data Analysis and Graphics Version.* Retrieved from `https://cran.r-project.org/doc/manuals/R-intro.pdf`

Verzani, J. (2002). *simpleR-Using R for introductory statistics.* Retrieved from `https://cran.r-project.org/doc/contrib/Verzani-SimpleR.pdf`

von Oertzen, T., Brandmaier, A., & Tsang, S. (2015). Structural equation modeling with Onyx. *Structural Equation Modeling: A Multidisciplinary Journal, 22*(1), 148–161.

Wang, W.-C., Chen, P.-H., & Cheng, Y.-Y. (2004). Improving measurement precision of test batteries using multidimensional item response models. *Psychological Methods, 9*(1), 116–136. doi: 10.1037/1082-989X.9.1.116

Warm, T. A. (1989). Weighted likelihood estimation of ability in item response theory. *Psychometrika, 54*(3), 427–450. doi: 10.1007/BF02294627

Weeks, J. P. (2010). plink: An R package for linking mixed-format tests using IRT-based methods. *Journal of Statistical Software, 35*(12), 1–33. Retrieved from `http://www.jstatsoft.org/v35/i12/`

Wickham, H. (2009). *ggplot2: Elegant graphics for data analysis.* Springer-Verlag New York.

Wickham, H. (2017). tidyverse: Easily install and load 'tidyverse' packages [Computer software manual]. Retrieved from `https://CRAN.R-project.org/package=tidyverse` (R package version 1.1.1)

Wickham, H., & Chang, W. (2017). devtools: Tools to make developing R packages easier [Computer software manual]. Retrieved from `https://CRAN.R-project.org/package=devtools` (R package version 1.13.3)

Wickham, H., & Grolemund, G. (2017). *R for data science: Import, tidy, transform, visualize, and model data* (1st ed.). Sebastopol, CA: O'Reilly Media Inc.

Yao, L., & Boughton, K. A. (2007). A multidimensional item response modeling approach for improving subscale proficiency estimation and classification. *Applied Psychological Measurement, 31*(2), 83–105. doi: 10.1177/0146621606291559

Yao, L., & Schwarz, R. D. (2006). A multidimensional partial credit model with associated item and test statistics: An application to mixed-format tests. *Applied Psychological Measurement, 30*(6), 469–492. doi: 10.1177/0146621605284537

Yen, W. M. (1981). Using simulation results to choose a latent trait model. *Applied Psychological Measurement, 5*(2), 245–262. doi: 10.1177/014662168100500212

Yen, W. M. (1984). Effects of local item dependence on the fit and equating performance of the three-parameter logistic model. *Applied Psychological Measurement, 8*(2), 125-145. doi: 10.1177/014662168400800201

Zeileis, A. (2005). CRAN task views. *R News, 5*(1), 39–40. Retrieved from https://CRAN.R-project.org/doc/Rnews/

Zhang, J., & Stout, W. (1999). The theoretical detect index of dimensionality and its application to approximate simple structure. *Psychometrika, 64*(2), 213–249. doi: 10.1007/BF02294536

Zimowski, M., Muraki, E., Mislevy, R. J., & Bock, R. D. (2002). *BILOG-MG: Multiplegroup IRT analysis and test maintenance for binary items.* Lincolnwood, IL: Scientific Software International.

Zumbo, B. D., & Thomas, D. R. (1997). *A measure of effect size for a model-based approach for studying DIF* (Tech. Rep.). Prince George, Canada: University of Northern British Columbia.

Zwinderman, A. H. (1991). A generalized Rasch model for manifest predictors. *Psychometrika, 56*(4), 589–600.

Index

analysis of variance (ANOVA), 56, 72

Bayesian statistics, 281–283
Bollen's plot, 80, 213
bootstrapping, 44–46

classical test theory, 31–56, 58,
    107–108, 280
cognitive diagnostic modeling, 280
computerized adaptive testing,
    278–280
Cook's distance, 215–217
correlation
    Pearson, 41, 47, 83
    point-biserial, 50
    polychoric, 91, 104
    tetrachoric, 41
CRAN task views, 277–278

data sets
    HSQ, 14–15
    SAPA, 41–45, 49–52, 91–92,
        114–141
    VerbAggWide, 159–160, 261–274
    efData, 60–66
    eirm, 196–210
    hcre, 237–242
    interest, 14–16, 25–27, 35–39,
        47–49, 77–91, 213–258
    mimic, 176–186
    multiplechoice, 52–53,
        163–166, 194–195
    negd, 243–246
    rse, 16–24, 145–157
    wiscsem, 93–105
    writing2, 69–72
    writing, 66–69, 72–73
    depression, 187–192

diagnostic plots, 212–222, 225–228
differential item functioning, 258–274
    item response theory likelihood
        ratio test, 269–274
    logistic regression, 264–269
    mantel-haenszel, 259–264

equating, 233–247, 280
    equate, 236–246
    concordance table, 239
    designs, 234–235
    equipercentile equating, 241–242
    equivalent groups, 237–242
    functions, 235–236
    identity functions, 237–241
    linear functions, 237–241
    linear Tucker equating, 243–246
    mean functions, 237–241
    methods, 235–236
    nonequivalent groups, 242–246
    nonlinear functions, 241–242
explanatory item response theory,
    193–210
    lme4, 196–210
    data structure, 194
    differential item functioning, 207
    interaction models, 206–210
    latent regression Rasch model,
        203–206
    likelihood ratio test, 209
    linear logistic test model,
        199–203
    long format, 194
    rasch model, 196–199
    uniform DIF, 208
    wide format, 194

factor analysis, 73, 75–105

common factor model, 75–76
confirmatory factor analysis, 76,
    92–105
  categorical data, 103–105
  continuous data, 93–103
exploratory factor analysis,
    76–92
  categorical data, 91–92
  continuous data, 77–91
  eigenvalues, 84
  estimation, 76
  factor scores, 89
  Kaiser's rule, 84
  parallel analysis, 84, 91
  rotation, 76, 88, 93
  scree plot, 84

generalizability theory, 55–73
  absolute error variance, 58, 65
  D study, 58
    one-facet, 64–66
    two-facet crossed, 68–69
    two-facet crossed with a fixed
      facet, 73
    two-facet partially nested,
      71–72
  dependability coefficient, 58–60,
    64–66, 68–69, 71–73
  design
    crossed, 57
    nested, 57
  facets, 56–57
    fixed facet, 56
    random facet, 56
  G study, 58, 60
    one-facet, 60–64
    two-facet crossed, 66–68
    two-facet crossed with a fixed
      facet, 72–73
    two-facet partially nested,
      69–71
  generalizability coefficient,
    58–60, 64–66, 68–69, 71–73
  relative error variance, 58, 65
  sources of variation, 57

units of measurement, 56, 57
universe-score variance, 57, 59,
    72

hierarchical linear models, 283–284

item analysis, 49–53
  distractor analysis, 52
  distractors, 52
  item difficulty, 49
  item discrimination, 49
  item discrimination index, 51
  item-reliability index, 51
  item-validity index, 52
item response theory, 105, 107–167
  mirt, 107–141, 143–166
  ability estimation, 128–133
  conditional standard error of
    measurement, 111–112,
    118–119
  dichotomous IRT models(, 107
  dichotomous IRT models), 142
  four-parameter logistic model,
    126–128
  generalized partial credit model,
    152–154
  graded response model, 154–157
  item characteristic curve,
    108–110, 116–117, 121,
    123–124, 126, 128
  item fit, 134–136
  item information function,
    110–112, 117–118
  local independence, 112
  model assumptions, 112–113
  model diagnostics, 133–139
  model selection, 139–141
  nested logit model, 160
  nominal response model,
    158–160
  one-parameter logistic model,
    113–119
  partial credit model, 144–148
  person fit, 136–139
  polytomous IRT models(, 143

polytomous IRT models), 167
Rasch model, 119–121
Rasch models for polytomous
    items, 144–151
rating scale model, 148–151
test information function,
    110–112
three-parameter logistic model,
    124–126, 228
two-parameter logistic model,
    122–124
unidimensional IRT, 107
unidimensionality, 112

linking, 280–281

Mahalanobis distance, 213–215
measurement invariance, 249–258
    configural, 251
    partial, 256–258
    strict, 253–256
    strong, 252–253
    weak, 251–252
multidimensional item response
        theory, 169–192
    multidimensional 2pl model, 184
    ability estimation, 181–184
    between-item models, 171–174
    bi-factor model, 189–192
    compensatory models, 170–171
    confirmatory models, 174–192
    cSEM plot, 181
    exploratory models, 174
    item characteristic surface, 171
    item contour plot, 179
    item information plot, 180
    mdiff, 175, 186
    mdisc, 175, 186
    multidimensional 2pl model, 175
    multidimensional graded
        response model, 186–189
    multidimensional Rasch model,
        184–186
    noncompensatory models,
        170–171

test information plot, 181
variance-covariance matrix, 186
within-item models, 171–174

path diagrams, 222–223
profile analysis, 284–285

Q–Q plot, 24, 217–218

R
    data manipulation, 16–21
    descriptive statistics, 21–22
    inferential statistics, 22–25
    installing R, 3–4
    masked functions, 13–14
    naming conventions, 9
    objects, 8
    overview, 1–29
    packages, 11–13, 27–29
    plotting, 24–27
    reading in data, 14–16
    reshaping data, 20–21
reliability, 41–46, 56, 58, 64–66, 68,
        69, 72, 73
    alternate forms, 41
    coefficient alpha, 44
    internal consistency, 41
    parallel forms, 41
    Spearman-Brown, 43
    split-half, 42
    test-retest, 41
RStudio, 2, 4–6

scales, 33
    categorical, 35
    continuous, 35
    interval, 34–35, 38–39
    nominal, 34–38
    ordinal, 34–38
    ratio, 34–35, 38–39
shiny, 223–231
standardized observational residuals,
        218, 220

validity, 47–49
    concurrent validity, 48

construct, 47
content validity ratio, 47
criterion-related validity
    evidence, 47
incremental validity, 48
predictive validity, 48

Printed in the United States
by Baker & Taylor Publisher Services